"Don't listen to anyone else's opinion of this book. Read it for yourself, and join the discussion about the future of our race. It might be the most important conversation you've ever had."

"This book is the most politically divisive book on climate change on the market today."

"Using historical events and facts, this author draws a future that is a chilling, all-to-real scenario."

"Buy it, read it, share it, and you will not regret it! It's time to actually participate in this conversation, and this book with give you the knowledge to understand what is happening to your world!"

"Robert Chapman takes complex topics and makes them understandable in a down to earth manner. The book is filled with laugh out loud moments, thought provoking conclusions, and compelling calls to action. It's a must read!"

"Thank goodness for this concise historical perspective to help predict our future, in a time of so many variables. Chapman will educate you in a down to earth manner. He makes history come alive, predicts a scary future, and then provides hope and solutions."

"First time author, Robert Chapman decided to write a book that needs to be purchased, read, and shared with everyone you know. This book is the cliff notes to the next human die off, that you will refer to over and over again!"

"This book will make you angry. The facts are undeniable and the logical conclusions that Mr. Chapman draws will make you question the very concepts of right and wrong."

Acknowledgements

No book is written by only one person, and there are a few people I'd like to thank. My family, who read the drafts objectively and suggested I publish under an assumed name (which I didn't do). To Mayra, my copy editor and muse. To Debb, who was the original inspiration to this book.

Finally, to Sara, a former employee of mine who quit in disgust and nearly ran me down in my own driveway after reading the first draft. If it hadn't been for you, I never would have finished this book. You let me know that my ideas are a catalyst for change. Wherever you are now, I hope that you are doing well.

The Next Human Die Off

(and how to prepare for it)

Robert Chapman

The History of Evolutionary Die Offs, Understanding Our Current Path, and Preparing for the Inevitable

----

Introduction

Chapter 1: A History of Human Die Offs – Disease, Famine, Wars, and Political Die Offs

Mass Deaths from War

Mass Deaths from Disease

Mass Deaths from Famine

Religious / Ideological Mass Deaths

Past Deaths vs Present Deaths

Chapter 2: How Mass Die Offs Benefit Humanity as a Whole

The Ability to Write History to Support Their Version of the Facts

The Ability to Pass Their Genetics Into the Evolutionary Timeline

The Ability to Pass Along Genetic Immunities and Higher Tolerance to Disease

The Ability to Pass Along Technological Knowledge

Reforestation After Mass Die Offs

Chapter 3: Governments and Class Structure – History and Predictions

Definition of the Development of Nations, Third World Countries, and Class Structure

Two Governments Overthrown by the People

The Futility of Armed Revolt in a Modern Developed Country

The Malthusian Catastrophe – Population Growth and the Availability of Food

Predictions of Death Rates Based on Social Class and Nation

A Quick Review

Chapter 4: The Most Likely Culprit – Climate Change and the Mismanagement of Natural Resources

Mass Death Scenarios

The Unvarnished Truth – Read This if You Don't Read Anything Else in This Book

Why Climate Change is (Still) Being Ignored

Artificial Intelligence – What Pure Logic Tells Us About the Coming Die Off

Chapter 5: The Last Hope - How to Slow Climate Change by Changing Population Levels

Population Changes from Conception to End of Life

Policy Changes that Reduce Population and Increase Conservation

The Green New Deal vs The New Deal of 1933

A New Conservation Corps?

Chapter 6: The Tipping Point – How it Begins

Scenario #1: The Perfect Storm of Rising Seas, Increasing Intensity of Storms, and our Crippling Lack of Immigration Reform

Scenario #2: An Outside Aggressor, Ammonium Nitrate, Social Media, and Deepfake Technology

Scenario #3: Wildfires, Food Production Pathogens, and Opening of Pandora's Box in the Arctic

Other Factors That Could Start the Tipping Point

A Footnote – Nuclear Detonations

Chapter 7: The First Year – What History Teaches Us to Expect

Effects of Starvation on the Human Body

The Spanish Flu Pandemic of 1918

Climate Refugees and What They Will Face

Chapter 8: Learning From History – Practical Applications

Predictions for the Future

Steps You Can Take Now

Keeping Your Loved Ones Safe During the First Year

Afterword

Appendix A – The Green New Deal

Appendix B – Books I've Read That Have Helped Me Write This Book

Introduction

Human die offs are as much a part of our evolutionary background as the loss of our tails, but most people remain blissfully ignorant of them. A new, future human die off is inevitable, and it's coming closer all the time.

Most people living in the developed world today have no real concept of mass human death. We've seen pictures of World War I and World War II, of course, as well as the images coming from the war zones of the Middle East and gang killings in Mexico. We've watched plenty of deaths on television and in movies. Occasionally, we've seen news stories about refugee camps in other countries, although any real suffering has been sanitized from the reports to make sure the sensibilities of the general public weren't affected.

However, a real understanding of mass death from famine, disease, or political circumstances is far beyond the experience of most people living in the developed world today. Europe and Russia are slightly closer to understanding mass death because WWII was fought there, but even so, most people with first-hand experience are in their 80's now. In the United States, the last war fought on US soil was the Civil War. Unless you are in the medical field, a first responder, actively in the military, or a veteran of a foreign conflict, chances are you've probably never even seen a body that wasn't previously embalmed. This is completely unique in history, and has really only happened during the beginning of the industrial revolution.

Ironically, it's part of our instinctual genetic makeup to expect death, both on an emotional level and a very real physical level. Who can argue that the love of a child isn't instinctual? Part of our evolution and the success of our species is to protect not only our biological children, but a strong emotional and evolutionary response to any child. On a physical level, we are all born with two primary fears – the fear of falling, and the fear of loud noises – but the body also has a complicated set of instructions when it senses danger, as well as other pre-programmed list of instructions, such as the mechanisms that employ our senses of taste and smell to detect whether food and water is safe to eat and drink.

The human brain has an almost infinite amount of these pre-programmed, genetic responses to a variety of stimuli. One of these situations is the fear, and the expectation of, mass human death by famine, war, or natural disasters. Why else do we watch movies with apocalyptic themes? Why else do we rubberneck at traffic accidents? Take one look at the ossuaries under Paris and find that we are fascinated by death as much as we are repulsed by it. It is in our genetic makeup to expect mass death. Collectively, subconsciously, we know that a mass die off is approaching.

If you are standing in line at the bookstore or using the "look inside" feature on Amazon, hopefully I've caught you with your cash still in your wallet. I want to be clear about what my intentions are in this book.

First – this is not a book on climate change. I do feel that the most likely culprit to the coming die off is climate change; and while I will quote some frankly terrifying information about the coming climate crisis, that's not what this book is.

Second – this book is not a "prepper" book. I'm not going to tell you what weapons to buy or how to booby trap your squash crop. I will be providing information that relates to preparing yourself for the mass die off, and paraphrasing some useful information from sources I have researched – but that's not the point of this book.

This book is a dispassionate look at past human die offs, and the significant benefits to the overall human race that became of these evolutionary events. We're going to discuss how the technological advances of the past fifty years have lulled us into a false sense of security that is going to make the coming die off even worse. We're going to discuss how mismanagement of our natural resources will make the deaths of millions of people inevitable.

Most importantly to the reader, we're going to talk about what you can do now to prepare, warning signs that the impending crisis is coming, the first year of the crisis, and how to survive with your loved ones safely.

In the second book of this series, we'll discuss restructuring after the die off, the convergence of ideas and technology, and the positive benefits that awaits us at the end of what will be the largest mass grave in history.

As I've been writing and researching this book, I've had to grow more dispassionate about the upcoming tragedy that is going to befall us. I'm sure that I will be criticized for this when this book goes to print. After all, it's one thing entirely to talk about 40 million people dying in the Mongolian wars 800 years ago, and another thing to talk about Bob and Linda down the street starving to death in 21st century America.

However, my self-imposed job isn't to mourn the passing of people – it's to prepare them and to look past this upcoming tragedy to a new future. Will the path to our ultimate salvation be littered with bodies? No question. Educate yourself, be prepared, and be ready. It IS coming and only the people who have prepared for it will survive.

Chapter 1: A History of Human Die Offs – Disease, Famine, Wars, and Political Die Offs

Before we can explore the future, we must talk about the past. In this first chapter, we'll talk about some of the worst die offs that have occurred during human history. We'll also introduce the four main causes of mass deaths. Don't worry – this won't take long, and we'll get to the good stuff soon.

Homo Sapiens has lived on the earth for about 100,000 years; however, the span of human recorded history is only the last 5,000 years. The first coherent writing ever found is Sumerian Cuneiform, a written language consisting of pictographs that originated in Mesopotamia, in modern-day Southern Iraq that originated approximately 3000 BC.

Mass Deaths from War

The first recorded war followed only 300 years later, between the Sumer culture and the Elamites. Of course, this wasn't the first war in history, only the first recorded. Jericho, arguably the world's oldest city, has archeological evidence of a solid fortification dated before 7000 BC. And a recent archeological find in Kenya of a mass grave of 27 humans, including six children and a late-term pregnant woman, were killed in a 10,000 year old turf war.

Nothing kills large numbers of people like war. In the ancient world, the fall of Rome took 8 million lives around the years of 476 AD, and the Mongol Conquest of Genghis Khan took a whopping 40 million lives between the late 12th and early 13th century. This number wouldn't be replicated in time of war until WWII, where an estimated 70 million people were killed.

Mass Deaths from Disease

Disease has been with us for our entire evolutionary existence. Diseases such as smallpox, malaria, typhus, consumption (or tuberculosis) and, of course, bubonic plague have alternatively defined and decimated our species as a whole.

In the ancient world, disease was common for many of the same reasons that it was in the pre-industrial age – that is, lack of clean water, and proximity of people to each other. Many times, disease and war would go hand in hand. For example, the Plague of Athens coincided with the Peloponnesian Wars in 430 BC. This plague, which has not been successfully attributed to a single disease, cost Athens tens of thousands of lives – some estimates reach as high as 25% of the population of Athens.

The worst pandemic in history, as far as deaths per capita, was the outbreak of bubonic plague in Europe in the mid-14th century. This plague, that lasted from 1347 to 1352, cost Europe 50 million people, between 50% and 60% of the population of Europe, and 7% of estimated world population. While Europeans (and by extension, Americans) consider this an exclusively European event, *yersinia pestis* has been around for millennia, first showing up in China in 225 BC, and covering the entire world in the years after the Black Death. Europe herself saw major outbreaks until well into the 16th century.

Mass Deaths from Famine

Famines have also plagued the human race for millennia. Some famines are caused by disease to crops, such as the Great

Potato Famine in Ireland from 1845 to 1853, that killed 1.5 million and caused another 2 million to flee the country – nearly 25% of the population of Ireland at the time. Some famines are caused by a natural disaster that destroys farmland, such as the Chinese Famine of 1907, where 25 million died due to flooding of farmlands.

The worst famines in history, however, have been a 'perfect storm' of political and environmental factors. Death tolls from these combination events have really only started to get astronomical in the last hundred years or so.

Stalin wasn't the first dictator to kill off his citizens in peacetime, but he was the most successful at it, at least up until 1938. The Soviet Famine (also called the Great Famine, or the Holodomor) was a combination event of three major factors – natural disasters that affected farming, a population explosion in the urban areas that increased the demand for food, and Stalin's polices of deliberately exporting grain from the Ukraine. Combined with The Great Terror, which was a campaign of political purges and murders of dissidents, Stalin managed to kill off somewhere between 6 to 8 million people between the years of 1932 to 1938.

But Stalin was an amateur compared to Chairman Mao in China. The Great Leap Forward from 1958 to 1962 was, in theory, a way to increase industrial output by utilizing socialist practices. Mao saw grain and steel production as the key pillars in industrial expansion, so he tasked farmers with doubling their output in crop yields (on the same land area), while in their spare time, building smelters to make steel – a task that most of them had no idea how to do.

Since crop yield was supposed to be doubled, the state collected the yields that the farmers originally were growing – which left nothing for the farmers themselves to eat. Of course, anyone who disagreed with Mao was labelled a 'dissident', so party members falsified reports of crop yields and steel manufacturing. In a very short time, the catastrophic results of Mao's policies were impossible to hide, and 43 million people died of starvation.

Religious / Ideological Mass Deaths

More people have died in the name of religion that for any other purpose in human history. Babylon and Jerusalem have been captured, sacked, and rebuilt more times than even historians can count. The Crusades, commissioned by the Vatican as a way of taking holy sites back from Muslim rule, killed anywhere between one million and three million people over 200 years. And there's more – lots more.

In 1517, when Martin Luther nailed his Ninety-Five Theses to the door of All Saints Church, he began the Reformation. One hundred years later, the 30 Years' War took 8 million lives in the conflict between the Papacy and the Protestants. Of course, the Vatican didn't stop at Muslims and Protestants – the Spanish Inquisition (Nobody expects the Spanish Inquisition!) saw Europeans burned, hanged, drowned, and tortured in inventive ways. Some estimates show the Vatican responsible for 100 million deaths between the 800 or so years between the beginning of the Crusades and the rise of Napoleon.

Politically motivated or ideological death tolls (also called politicides) deal with genocides not originating from war. These events are almost exclusively a 20th century phenomenon, and they happened when political leaders just went in and started killing people. In most cases, the people being killed were labeled either as political dissidents, or just happened to be on the wrong side of the more powerful group. These include the Biafra genocide between 1967 and 1970 (death toll, 2 million), the Rwandan Genocide in 1994 (death toll, 1 million) Pol Pot's Killing Fields in Cambodia during the years of 1975 and 1979 (death toll, 2 million), and Idi Amin's brutal rule in Uganda from 1971 to 1979 (death toll, around 300,000).

Past Deaths vs Present Deaths

While the average person living in a developed country will certainly agree that these are all tragedies, not many of them will have any real feelings about these events one way or the other. They are all history now – just pictures in a book - with the affected people buried years ago.

But that's not really true, is it? Depending on where you live today, these events shaped your history, and the history of those around you. They affected the lives of your ancestors, who survived these events – or you wouldn't be here, reading this book. It's the story of the many servicemen that survived WWI but died from the Spanish Flu instead. It's the story of the men who fought in WWII and lived to come home and begin families, and the many that didn't and became a white marker in a cemetery. It's the men and women who faced enormous odds and an uncertain future to flee Ireland in 1845, or stayed there to die. It's the story of the Jews who saw the writing on the wall in 1938 and left everything but the clothes on their backs, and those that stayed and lost everything, period.

Sometimes, it's the story of the lucky versus the unlucky, or the story of people with means versus people without them. In rare cases, it's the story of those who were prepared, who saw the warning signs, and were able to take the necessary steps to prevent a global tragedy from becoming a personal one.

In the coming chapters, I'm going to show you why we are facing a mass die off of an unprecedented scale. I'm also going to make a valid argument that mass die offs are a necessary part of human evolution; that, in fact, we are overdue for one. Because we are overdue, and because of our population numbers, it will be the largest mass grave in history.

Here's the point of this chapter, and maybe this book. In one hundred years, or five hundred, or a thousand – your progeny should live with the knowledge that you survived this mass die off. They should flourish and multiply past this coming tragedy and into the future. The alternative is that you are one of the millions that die, and that you become just another picture in a book.

Chapter 2: How Mass Die Offs Benefit Humanity as a Whole

"To the victor go the spoils." This proverb in its present form was first uttered in 1832 by a New York Senator named William Marcy. The actual quote could have come from a political exchange today: "They (Democrats) see nothing wrong in the rule that to the victor belong the spoils of the enemy." Bi-partisanship at its finest!

The Senator was more correct than he knew, especially for living in an age that had no concept of the human genome. The individual that triumphs over any conflict, whether it be war, famine, or disease, has the opportunity to change the future for them, and for their progeny, in a variety of ways.

The Ability to Write History to Support Their Version of the Facts

Let's dive into the most morally ambiguous statement first, shall we? The victor of any conflict gets to write the history books. Imagine for a moment if the Axis powers had won WWII (and they probably would have, if Hitler hadn't decided to fight a two-front war). The Third Reich was directly responsible for the death of 6 million Jews, as well as millions of gay men, blacks, Soviet citizens and POWs, and ethnic Poles. Hitler also nearly extinguished the Romanian Gypsy population. Nearly 17 million people died as a direct result of Nazi persecution.

However, in what is now the United States, my ancestors engineered a successful genocide of the Native American people. It's unknown how many Native Americans lived in what is now the United States when Columbus "discovered" America in 1492, or when the Mayflower landed at Plymouth Rock in 1620. Estimates range from a low of 2.1 million to a high of 18 million. We do know that by 1800, long before the enforced colonization of Native American children through government funded boarding schools, long before Custer, and long before the Trail of Tears, the Native American population had dropped to 600,000. That's somewhere between 72% and 97% of the overall indigenous population of the United States killed.

There is no question that Hitler was evil and misguided. But, arguably, so were the ancestors of most people living in the US today. It's a little-known fact that Hitler actually modelled the concentration camps of the Third Reich after American concentration camps for Native Americans. So why is Hitler the litmus test of every evil impulse we have as a species, and the "Founding Fathers" get a free pass? Why do we celebrate Columbus Day and Thanksgiving? Simple – the US government won. It's the US that gets to write the history books.

Imagine if Britain had won the Revolutionary War. "A group of traitors to the Crown tried, and failed, to establish an independent government in the New World." Of course, Britain was destined to lose its grip on the American colonies at some point. Geographically, the New World was just too far away to govern adequately. But imagine how history would have been changed if the US had lost the Revolutionary War – and what the history books would have said about Washington on the Potomac.

While this is the most morally ambiguous "spoils", it still has an undeniable impact on what the world is today. Without the conflicts our ancestors fought; the technology of today, our political systems, down to the very food we eat would have changed. How did these wars benefit humanity as a whole? I suppose that's why it's morally ambiguous, especially considering the impact of the coming die off. The easy answer is that the mass death of an existing population in an armed conflict allows a surplus of resources for the winners. The end of WWII saw the largest population increase in the history of the United States, and quite possibly the world – the Baby Boomers. The mass death of the American Indians paved the way for Europeans to colonize the New World. But even if you don't believe that the impacts of the conflicts brought us to a 'good' place in our history, it's undeniable that it's a true place. After all, you're here, and so am I – right?

The Ability to Pass Their Genetics Into the Evolutionary Timeline

On an individual genetic level, the winners of an armed conflict not only get to write the history books, they also get the advantages of populating the land. One of the best examples of this concept is the modern-day areas of Asia that were a part of the Mongol Empire, where it's estimated that 16 million people living today are direct descendants of one man - Genghis Khan.

Here's an even better example. After the fall of Rome, the population of Europe shifted dramatically. There's even a term for it – it's called the Migration Period, and historians believe it occurred between 375 and ending somewhere in between 700 and 800 AD. In this period, what was the Western Roman Empire (read – most of modern-day Europe) was forcefully re-taken by the Huns, the Lombards, and a Germanic tribe called the Franks – who are the predecessors to the population of modern-day France. Historians still debate on whether the fall of the Western Roman Empire was a result of the Migration Period, or whether it was caused by it. But for our conversation, it doesn't really matter – this migration period has a direct result on pretty much any person with any English, French, Slavic, or Germanic ancestry currently living in the US – 62% of Americans.

But we can go even farther back. Every person living in the world today has anywhere between 2% and 4% percent of Neanderthal information in their genetics. In this case, the exception proves the rule - the Neanderthals died out over 40,000 years ago, but traces of them still exist in our very bloodstream. What hard-fought but long forgotten battles must have happened in that pre-historic age? What if some far-distant ancestor had died in battle instead of living to become a part of your parentage? Would you still be here, still recognizable, but with a different shape to your eyes? Or would you wink out of existence?

In the pre-industrial world, the winning side passed a superior set of genetics on to their progeny. When you're fighting on horseback with a bow and arrow, swords, or other weapons of hand-to hand combat, the winning side has better eyesight, reflexes, physical strength – which are pretty admirable traits to be passed along in the gene pool. These traits, passed down through generations of population and migration, have benefitted humanity as a whole. Much like the ability to put their spin on history, the victors of armed conflicts through the ages have brought us, genetically, to where we are today.

The Ability to Pass Along Genetic Immunities and Higher Tolerance to Disease

The first two points deal with armed conflicts, but the survivors of pandemics have played a huge role through history in strengthening our genetics. Again, there are several examples of this, but we'll talk about two here – smallpox and plague.

Cortez and Smallpox in Central America

It's hard to comprehend the truly devasting effect that smallpox had on the New World. The indigenous Aztecs in central

America, and native Indian tribes north of the Rio Grande had no immunities to the disease, and no idea how to treat it. Moreso, Europeans had developed immunities without knowing it, because of their exposure to other, milder forms of the disease, such as cowpox and horsepox.

In Central America, a Spanish explorer and conquistador named Hernan Cortez landed on the coast of modern-day Mexico in 1519 and claimed the land for Spain. The Aztec Empire was arguably waning, mainly because of their religious beliefs – it's hard to maintain a working class when your religion is based on human sacrifice, and the Aztecs were really, really good at human sacrifice. Regardless, smallpox killed around 25% of the indigenous population in less than 5 years, including the Aztec emperor, Cuitláhuac, and many of the leaders of the Aztec army. So many people died that there was no one left to bury them, and the Aztecs would simply demolish a smallpox victims house, leaving the bodies to rot under the rubble.

However, 600 years after Cortez and the fall of the Aztecs, a whopping 21% of Mexican nationals identify as being indigenous. That's compared with 2% of the American population identifying as Native American, and about 5% of Canada's population identifying as Aboriginal. It's been said by many genetics scientists that the population of Mexico is the most genetically diverse population of any nation on Earth.

So why did the indigenous people fare so well in Mexico, and so poorly in what is now the United States and Canada? A good portion, is, of course, the ideology of the conquerors. The Spanish conquistadors believed in spreading the word of Christ and converting the indigenous people to Catholicism. Farther north, the new, largely protestant new Americans believed that the land was literally theirs by divine right, and that the indigenous people were inferior BECAUSE they didn't believe in God. While both concepts are rooted in Christianity, there's a big difference between conversion – even conversion at sword point - and genocide.

But the better reason is that Cortez intermarried with the indigenous people, and by proxy, encouraged his men to do so. In fact, his first wife, Dona Maria, was originally from the Mexican coast, and served as his mistress, consort, and interpreter through much of his conquest. As you can imagine, her name brings a mixed response in Mexican culture – some see her as a traitor to her people, some as an unwilling hostage, and some see her as the quintessential mother to the new Mexico. Regardless, her impact is undeniable – her child with Cortez was one of the first Mestizos (a person of mixed European and indigenous heritage). In modern Mexico, 86% of the population share mixed blood of European and indigenous peoples. It's also worth pointing out that this number is based on genetic research, and the 21% of Mexican nationals who identify as indigenous are based not on genetics, but by their own oral family history.

Because Cortez shared this genetic heritage with the indigenous people, and his men did as well, the resulting generations were genetically stronger and more resistant to the new European diseases. Combined with the overabundance of resources because of the smallpox epidemic, resulting generations of mixed ancestry fared much better than either of the two groups, independently, would have done.

The Black Death in Europe

Unlike Cortez, who brought smallpox as a conqueror, The Black Death was brought to Europe in trade ships from Asia. As plague is a zoonotic disease, it arrived in Europe literally on the backs of the black rats that infected the trade ships – in the form of fleas.

Here's how the disease is transmitted to humans: A flea bites an infected rat and becomes infected itself. Once the flea is infected, the bacterium multiplies in the flea's digestive track. The multiplying bacterium creates a blockage that hinders digestion, causing the fleas to starve. The flea will literally vomit the infected blood up to try and clear the blockage. So – a starving, infected flea will bite the rat, and then vomit infected blood back into the bite site, which, in turn, infects the rat. The rat goes on the be bitten by other fleas, then dies of the disease. The fleas, now infected, go on to find the nearest living host, which sometimes will be human. It's really a fascinating disease, although probably not when you are watching your lymph nodes swell to the size of tennis balls, and patches of your skin turning black while hallucinating from high fever.

Again, it's hard to overestimate the impact that the Black Plague had on 14th century Europe. A conservative estimate shows half of the population of Europe dying in the four years of the Black Plague. There are some reports in France of 90% of the population of some towns dying. Most people in the modern age believe there was only one outbreak, but that's inaccurate – plague swept Europe, Asia, and Africa almost generationally for hundreds of years after the Black Death in 1348. It's last major outbreak was in 1665, nearly 300 years later, in the Great Plague of London, which killed nearly one fifth of the city's inhabitants.

But wait – read that again. The initial outbreak of plague in 1348 killed 50% of the population of the continent, and the outbreak 300 years later only killed less than 20% of the population of London. While 20% isn't great odds, especially considering the horrible death that plague brings, I'll take it any day over half. What changed the percentage so drastically?

There are a few factors. The first factor is, ironically enough, cats. Superstition ran high in 14th century Europe, and cats were seen as hosts for demons or witches – a superstition that persists to this day in "unlucky" black cats. Laws were enacted making it illegal to own a cat – which, of course, caused the rat population to go unchecked. It took several hundred years for people to realize that their perceived enemy was actually one of the few forces that kept the disease in check.

The second factor was government quarantine. Although enforcement was spotty at best, and sometimes even used as a political punishment, quarantine was effective in checking the spread of the disease in later outbreaks. Additionally, once the poorer sections of a population started showing an outbreak, the wealthiest classes were able to flee the infection before it began spreading, increasing their chance of survival. It's specifically noted that the nobility suddenly decided, en masse, to visit their estates in the country in the spring of 1665 – the beginning of the Great Plague of London. They were followed by the merchants and lawyers. Even the clergy decided they'd be better off ministering to their flocks that didn't reside anywhere near London.

From the standpoint of this book, however, the third factor is the most interesting one. The population of London in 1665 had a genetic survival advantage over the population of Europe in 1348. The majority of people who survived the plague (and would go on to have children) had a genetically stronger immune system than people who did not survive. This example of positive selection allowed people with a predetermined genetic disposition to survive and multiply where other people who didn't have this mutation did not. It's not that the plague changed the human genome – it's that people who already had a mutation for a stronger immune system were able to fight the infection better. These survivors passed along that mutation to their children.

Finally, after each cycle of the plague, the resulting availability of resources made the survivors even healthier. By having access to an abundance of better food and cleaner living conditions the overall population became stronger, healthier, and therefore, better able to fight the next wave of the disease.

The Ability to Pass Along Technological Knowledge

In the ancient age, the ability to use and to pass along advanced technologies was an instrumental part of victory in an armed conflict. For example, the first group of people to make steel was the Hittites, sometime around 1400 BC. Their use of steel gave them a distinct advantage over their enemies, which, in turn, had a huge effect on the balance of power in what is now Turkey.

However, technologies used in warfare were, and are, only a part of the advantages that the winning side passed along to their progeny. A good example of this is the series of Roman aqueducts that were built throughout the Roman Empire starting in 312 BC. At the height of their use, there were over 1000 aqueducts, supplying water to dozens of major cities, and moving millions of gallons of fresh water to populated areas per day. This technology was only one of the reasons for the successful reign of the Roman Empire for over 800 years, and it had nothing to do with war, but rather, civil engineering.

There's a big difference between war and warfare. War is defined by a state of conflict between two groups. Warfare deals with the technology and strategy of war. For centuries, there was a balance between offensive and defensive measures. A better arrow necessitated better shields on the opposing side. Trebuchets and other siege engines necessitated higher, stronger walls. Better metallurgy required better armor. The winning side of any conflict was able to pass along these technological advancements to the benefit of their cause and their progeny.

This began to change around WWI, when widespread use of chemical agents like mustard gas found its way into modern warfare. With chemical or biological agents, you don't have to have a superior fighting force or better weapons. The Geneva Protocol of 1925 recognized this, and outlawed chemical and biological weapons for use in war.

Of course, the technology of warfare changed again in WWII, about the same time the ambient air temperature of Hiroshima reached 9000 degrees Fahrenheit. With the advent of the hydrogen bomb, you didn't have to have a better army, military tactics, or weapons – you just had to have nuclear technology and the ability to use it.

So, how have the mass deaths in warfare benefitted the human race today? Our ability to pass along technical knowledge in the modern world is certainly a part of this – the advances of modern medicine, agriculture, and communications is as much about the wars fought and won as it is about technological advances that caused the winning side to proliferate.

But as military strategy changed in the last hundred years, part of the knowledge that we have learned and have (hopefully) passed along to our progeny is when NOT to use these technologies of war. Is it possible that man is at a turning point in our evolution where we realize that the horrors of war are not worth the cost? Is it possible that part of the knowledge that we gained from Hiroshima and Nagasaki is that there are no winners in warfare – at least not at the nuclear level?

If so, then we truly have passed valuable knowledge on to our children. The survivors of the coming mass die off will have access to advances in medicine, agriculture, and communications unheard of in human history. They will also inherit the technologies of warfare, including our nuclear arsenal. Maybe, just maybe, they will also inherit the wisdom to not use it.

Reforestation After Mass Die Offs

It's funny to think that pre-industrial man left a carbon footprint, since it seems like 'carbon footprint' is the newest buzzword of climate change. Of course, it's true - since the dawn of time, men have quarried stone for buildings, homes and tools, cleared lands for crops, and raised livestock. They built their homes and used firewood to heat them.

Gold was the first metal to be mined and used for art and jewelry, sometime around 9000 BC. However, true smelting of ores didn't really happen until the beginning of the Bronze Age around 2300 BC. The mining, smelting, and weapons production of these metals and alloys all contributed to early mans carbon footprint – all the way up to the invention of the steam engine in 1698 (which, ironically, was designed to help pump water out of coal mines).

But we really didn't get started with burning fossil fuels until the beginning of the Industrial Revolution around 1760. Metallurgy had finally reached a point where quality steel was available for machination, and coal was being mined at a steady pace. In the United States, Eli Whitney invented the Cotton Gin in 1793, which forever changed the American South, and the world. In 1859, the first American oil well was drilled in Pennsylvania. In the first five years of the 1900's, the Spindletop oil well was drilled near Beaumont, Henry Ford's first mass produced car rolled off the assembly line, and Orville and Wilbur Wright made history at Kitty Hawk.

We've already talked about Genghis Khan (and will again in this book) but his legacy is undeniable. One of the unintended consequences of the Mongol conquest was an actual lowering of carbon in the atmosphere because of re-forestation. Julia Pongratz, a Research Scientist for the Carnegie Institution, writes:

'We found that during the short events such as the Black Death and the Ming Dynasty collapse, the forest re-growth wasn't enough to overcome the emissions from decaying material in the soil. But during the longer-lasting ones like the Mongol invasion... there was enough time for the forests to re-grow and absorb significant amounts of carbon.'

The Mongols were responsible for the deaths of about 40 million people – about 10% of the population of the world in his time – starting at his coronation in 1206, and ending not with the death of Genghis Kahn, but with the final expansion of the Mongol empire in 1279. In comparison, the Black Death killed about 7% of world population, the Spanish Flu Pandemic of 1918 killed about 3% of world population, and WWII killed about 2% of world population.

Let's look at this another way. In 1950, these were 2.55 billion people living in the world. At the time of this writing, there are 7.53 billion people. In the seventy years it took for the Mongols to conquer Asia and kill 40 million people, we have increased world population by nearly 5 billion people.

The simple fact of the matter is that the modern world has too many people for the earth to support. The earth's resources are finite, and our modern society is using them too fast for the earth to heal and replenish itself. It's simply unsustainable.

Before anyone accuses me of looking forward to the deaths of millions of people – I'm not. I don't want a die off to occur. I'm saying it's the inevitable conclusion of the population explosion we have witnessed in the last 100 years. Because of our overpopulation, the scarcity of resources will force a mass die off. We are potentially looking at the deaths of millions of people to balance the scales between population and the natural resources of our planet.

During the time of the Mongols, the earth was able to heal itself through reforestation. It's been speculated that during the die off of the North American indigenous peoples between 1492 and 1800, the planet saw a similar cooling period because of the reduction of the indigenous populations of what is now Mexico and the United States. What will it take for the earth to once again be able to heal?

Imagine a 10% reduction of the world's population today – that means that over 750 million people would die. That's more than the population of the entire North American continent. Now imagine a reduction in population to the same level as the world's population in 1300 AD - estimated at approximately 400 million people. That means 7.13 billion people would die. That's 94% of current world population.

Will the coming die off be a tragedy of an unprecedented scale? Absolutely. However, as history has shown us in all mass die offs, the remaining survivors will benefit. They will have better genetics, a biosphere with less atmospheric carbon, more resources to share, and a base of knowledge unlike any other in history.

"AND I LOOKED, AND BEHOLD A PALE HORSE: AND HIS NAME THAT SAT ON HIM WAS DEATH, AND HELL FOLLOWED WITH HIM. AND POWER WAS GIVEN UNTO THEM OVER THE FOURTH PART OF THE EARTH, TO KILL WITH SWORD, AND WITH HUNGER, AND WITH DEATH, AND WITH THE BEASTS OF THE EARTH." REVELATIONS 6:8 KJV

Chapter 3: Governments and Class Structure – History and Predictions

Definition of the Development of Nations, Third World Countries, and Class Structure

Let's step back for a minute and discuss some terminology that is going to be important for this chapter and the rest of the book. During the Cold War, the world nations were divided into three groups: The First World countries, who were either democracies or countries that were allies of other democracies; the Second World countries, who were aligned with Communism and the Soviet Union; and the Third World countries, who didn't really pick a side.

After the Cold War ended, these terms changed. Now, scientists and policymakers look at different divisions: Developed Nations, which are counties that have a strong industrial, manufacturing, and military presence; and Developing Nations, which are countries that are still trying to develop their industrial presence, and more important, to stabilize their class structure. The term "Third World" is still used, although it now means countries that have exceptionally low literacy rates and exceptionally high infant mortality rates. These countries have an unstable economy because there is no middle class. Without this key component, the poor don't have any way to escape poverty as there is no next step up the financial ladder. The wealthy and/or ruling class keep all the money and resources to themselves, further destabilizing the economy and allowing poverty to run rampant.

We are also going to talk about class structure in this chapter. A quick note on social classes – most sociologists (a scientist that studies how society works) believe there are five basic classes in the modern world. There are, in order: The Wealthy / Ruling class, Upper Middle Class, Lower Middle Class, the Working Class, and the Poor.

Two Governments Overthrown by the People

For centuries, political power was kept literally "in the family". In Europe particularly, a father would pass rule to his son. His son would marry a member of the family – typically a close cousin, and the reign would continue. The wealthy or ruling class of any society was changed in one of two ways: an armed conflict with an enemy, or a revolt within the previous ruling structure. There are hundreds of examples of this, but we are going to focus on two of them – the collapse of the Ming Dynasty, and the French Revolution.

The fall of the Ming Dynasty occurred around 1640. The Ming ruling class lost their hold on China for several reasons – both natural and political. Economically, the Ming ruling class relied on a silver standard. Due to world events dealing mainly with piracy in the early part of the 16th century, silver suddenly became less attainable, causing a significant drop in the value of Chinese currency of the time. Coupled with a series of natural disasters that, in turn, triggered widespread famine and disease, the ruling class was faced with serious opposition from both the middle classes (who had lost most of their wealth in the devaluation of the currency) and the rural working classes. Armed rebellions and peasant uprisings were common, and during the 60 years that it took to shift the balance of power to the Qing Dynasty, over 30 million people were killed.

The French Revolution is probably more relatable to Europeans, and by proxy, Americans. It's certainly more familiar, at least as far as name recognition. This is probably due more to *"Les Miserables"* than actual history classes – but I digress.

The French Revolution marked a shift of power from the ruling class of the Monarchy, to a political republic, and finally, a military ruling class. This happened largely because of the growing division between the wealthy and the poor – the "haves" versus the "have-nots". While the body count was comparatively low at only 40,000 people, most of those killed were the ruling class of the old feudal system, most notably, of course, being King Louis XVI and Marie Antoinette.

Due to King Louie's poor financial policies, including a poorly planned allegiance with the new United States, France was on the brink of bankruptcy. Combined with several years of poor crops and cattle disease, the urban poor and the rural working classes were fed up with the Monarchy. Ironically, King Louie set up a meeting between France's clergy, nobility, and shrinking middle class to try and discuss their grievances. By then, however, it was too late – the nobility didn't want to give up their status in the feudal system, and the wealthy were not interested in taxing their wealth to provide relief to the starving poor.

The meeting set by the King did convene on May 6, 1789, but it was hopelessly deadlocked. Although the King did capitulate his power to a general political assembly on June 27 of that year, it wasn't enough to placate the people. A group of rioters stormed the Bastille prison on July 14, 1789 and began the French Revolution. In June of 1791 the King tried to escape the country and failed. A group of extremists attacked the royal residence in Paris and arrested the King and Marie Antionette on August 10, 1792.   They were both put to death by guillotine – King Louis on January 21, 1793, and Marie Antionette just nine months later.

Following their deaths, a 10 month wave of arrests and executions of 'enemies of the Revolution' were carried out by the Committee of Public Safety – a period of time known as the Reign of Terror. The director of the Committee of Public Safety, Maximillian Robespierre, was himself executed in 1794. After a series of failed legislatives, a young general named Napoleon Bonaparte took power in a military coup. His rise to power would eventually see the domination of France over much of modern Europe, and would end with his failed attempt to invade Russia. This particular mistake would also cost another military leader his rule, and the Reich, nearly 160 years later.

The Futility of Armed Revolt in a Modern Developed Country

Both of these shifts of power occurred as a direct result of the poor and the working class uprising against their leadership. However, there have been some pretty significant advances in weapons, armament, and security technologies since the French Revolution. It would be incredibly difficult for a modern leader in a developed nation to have his or her power overthrown in an uprising, and really has been since an assassin's bullet changed history in Dallas in 1963.

Can you imagine an armed group entering the White House and arresting the President of the United States? The US spends a whopping 54% of all federal spending on the military. That's nearly $600 BILLION a year. How is a group of armed, working class citizens going to wage any kind of offensive against the armament that kind of money can buy? It would be almost impossible. A military coup seems more likely, quite possibly within the Secret Service itself.

Still, can you imagine a modern-day Napoleon taking over the United States Armed Forces? What would it take to overthrow the leadership of Japan in the modern age, or any of the countries in the EU? And how would a military ally from a sister nation treat a modern-day coup in the United States, or Britain, or even Mexico?

The fact of the matter is that the governments of developed nations are now so large and so powerful that they cannot be overthrown by the people. Because of this, modern lawmakers are completely insulated from having to make policy changes for the people. We'll talk more about this in Chapter 4, but the fact of the matter is that it will take nothing less than a mass die off to affect any real change in the balance of power in developed countries of the world today.

However, in the Third World, overthrow of unstable governments take place all the time, and will continue to do so as we come closer and closer to the mass die off. Countries like Sudan (who has been in the grip of two civil wars in the last fifty years), Venezuela (who could've been a lot closer to a Developed Nation rating, if not for the meddling by the US government – but that's another story), Somalia, and other countries in Equatorial Africa and South America are in far more danger to uprisings and their leaders being overthrown than Developed Nations.

The Malthusian Catastrophe – Population Growth and the Availability of Food

Nearly 225 years ago, an English economist named Robert Thomas Malthus wrote about population and food availability. His theories revolved around concerns of food production, food abundance, and how these factors affected population growth.

Malthus believed that food abundance would never be used to increase quality of life for a society – rather, that food abundance would always fuel population growth. His published work, "An Essay on the Principle of Population", also argued that food production acted as a balance to population growth. His theory was that humans, as a species, would increase their population until the lower classes suffered from food deprivation and malnutrition.

In other words, humans would never use food abundance to provide for a better society – food abundance would be used to increase population to the point that the lower classes would become expendable. In his writings, he may have been the first predictor of the next human die off. From our friends at Wikipedia:

"The main point of his essay was that population multiplies geometrically and food arithmetically; therefore, whenever the food supply increases, population will rapidly grow to eliminate the abundance. Thus eventually, in the future, there wouldn't be enough food for the whole of humanity to consume and people would starve. Until that point, however, the more food made available, the more the population would increase. He also stated that there was a fight for survival amongst humans, and that only the strong who could attain food and other needs would survive..."

Malthus wrote and published his essay in 1798. He never could have envisioned the world we live in today. The "Malthusian Catastrophe" is a term used to describe the worst-case scenario described by his essay. However, the sheer numbers of population growth since 1798 have increased his theorized catastrophe to an amount of people beyond the entire population of the earth living during his lifetime. I think that, if Malthus lived today, he would join the number of scientists terrified for the future of humanity. He understood that other factors come into play when discussing food abundance and population growth. However, he never could have predicted what climate change would do to food production.

Malthus would have been familiar with the term "decimate". It's a Roman term that literally means to "kill by a factor of 1 person in 10". The population of the earth in Malthus's time was 901 million. In our time, as we've discussed, it's 7.7 billion. A decimation of the earth's population would be 770 million people. And frankly, I believe that we are looking at significantly more than a decimation of the earth's population. I believe that the next human die of will see a minimum 40% reduction of the entire population of our planet, and possibly as high as 75%.

Predictions of Death Rates Based on Social Class and Nation

In the coming human die off and the years leading up to it, the division between social classes will become significantly wider and more visible that it is today. A recent survey reported that 52% of Americans are considered middle class. The official poverty rate – that is, Americans living below the poverty line – was 14.5% of the American population in 2013. One in eight people in the United States is on some form of government assistance.

Humans are exceptionally adaptable creatures, but we have reached a point in our evolution that is truly unprecedented. We have so much more than we can even use. In developed nations, we waste almost more food now than we eat. In the United States, we waste 50% of all produce grown in our borders. Even the poor in developed countries have a level of education, food availability, and healthcare simply unheard of – completely unfathomable - even 100 years ago.

The most basic understanding of statistics shows that these levels are unsustainable. The law of averages says that we must have downturn for every upturn. The downturn in this case is going to be a very large number of starving people. However, the tragedy humanity will face in short term and the long term are going to be very different for different people, and the factors affecting your survivability will be based on social class and skill level.

If you are in the ruling class of a developed nation or you are wealthy, chances are you'll be just fine, at least in the short term. Even in times of extreme poverty by the rest of the population, such as the Great Depression, the wealthy don't miss that many meals. If you've got the money to provide security for yourself, medical care for you and your family, and the money to pay the high prices for food, you shouldn't worry about the short term. The long term, however, may be troublesome….

Underground Bunkers – How the Ruling Class Will Retreat During the Die Off

If you've never read "The Time Machine" by H. G. Wells, you really should. Better yet, see the 2002 movie of the same name. The movie added some details that were lacking in the book. I won't give away any spoilers, but the premise of the book is a man that builds a time machine to find that, in the future, the human race has split into two distinct species – the Morlocks, a technologically advanced race who live underground, and the Eloi, a childlike, vegetarian race that lives above ground.

I've always wondered about the Morlocks. H. G. Wells wrote the book in 1895, long before the invention of the hydrogen bomb. In our time, of course, we know about the threat of nuclear war, although most people don't really comprehend the finality of it. Even a limited exchange would cause a feedback mechanism that would have huge impacts on the rest of the planet.

Imagine a limited exchange between India and Pakistan. Each side lobs 50 weapons at each other, for a total of 100 nuclear explosions. They certainly have the capability of doing so – both India and Pakistan have somewhere between 130 to 160 nuclear weapons in their respective arsenals. Doesn't sound like much, does it? There have been over a thousand nuclear weapons tests worldwide since 1945.

The problem is the consumption of oxygen in the affected areas. Most nuclear tests occurred in deserts, underground, at sea, and even in space. A nuclear weapon over a city creates a firestorm that burns hot enough to turn anything into thick, black carbon smoke. I'm talking about an entire city of fire, where even concrete gets hot enough to burn, where square miles are releasing toxic smoke. This type of fire changes not only the balance of oxygen in the area, but also changes oxygen in the ozone layer. It also has a worldwide effect on breathable oxygen in the atmosphere because of the amount of oxygen used in the burning of a city.

No one knows for sure, of course, but it's theorized that even the limited exchange described above would be enough to reduce oxygen levels to the point where the air would be unbreathable anywhere in the world. And there is no question that a full out nuclear exchange between any of the world's superpowers would render the entire earth uninhabitable by several factors – the most notable of which isn't the extinction of all plant and animal life, but the complete removal of breathable oxygen from our atmosphere – worldwide.

The wealthy and ruling class know this, of course, and have created huge bunkers to retreat to in the case of attack. However, you've got to wonder – if the people that were responsible for billions of deaths were to go underground, how long could they live there? If the surface of the earth was incapable of supporting life, how long could you stay underground? The exact amount of supplies these bunkers store is, of course, classified. The largest of them certainly have hydroponic systems setup to grow their own food. But everything is finite. Could you supply 25,000 people underground with enough food and water for 200 years? 300 years? Maybe H.G. Wells knew something we didn't?

At any rate, it's a relatively well-known secret what the wealthy and ruling class will do in the case of a nuclear attack. What is less known is the number of bunkers, the size of them, and the planned uses for them as we get closer and closer to the mass die off.

Let's address the planned uses first. The most well-known bunker in the United States is the Cheyenne Mountain Complex in Colorado. This military complex houses the equivalent of 15 three story buildings under 2000 feet of granite. Not only is it fed freshwater from underground springs, it's also got 6 million gallons of fresh water in reserve pools, and 510,000 gallons of diesel that powers a 10 gigawatt series of generators and battery backups. It's designed to deflect a 30 megaton nuclear explosion from 2 kilometers away. To put it mildly – it's secure. From the standpoint of surviving a nuclear war, it's where you'd want to be.

However, what other threats could cause the facility to close it's blast doors and seal itself from the outside world? The answer might surprise you. According to our friends at Wikipedia: *The threats, in descending order of likelihood, that the complex may face are "medical emergencies, natural disasters, civil disorder,*

*a conventional attack, an electromagnetic pulse attack, a cyber
or information attack, chemical or biological or radiological
attack, an improvised nuclear attack, a limited nuclear attack,
[and] a general nuclear attack."*

Would the American government seal up the Cheyenne
Mountain Complex in case of a civil emergency? Yes, they
would, and already have – the last time the facility was fully
sealed was the terrorist attacks of September 11, 2001. What
other instances would the government seal the facility? A
global pandemic? Civil unrest in the wake of an economic
collapse? What about a widespread natural disaster, such as a
volcanic eruption in a populated area? Yes, Yes, and Yes.
Although these facilities were envisioned during the cold war
and built under the threat of nuclear war, they are also there to
protect the ruling and wealthy classes from any foreseeable
threat. In other words – they close the doors and let the rest of
the country live or die as circumstances allow. When it's safe,
they come back out again.

The Cheyenne Mountain Complex is the best known, and maybe
the largest of several bunkers in the United States. No one
except the government really knows for sure, and when asked,
the government has politely told pretty much everyone to go
have sex with a pineapple. Some of the more public bunkers
include the Presidential Emergency Operations Center
underneath the East Wing of the White House, the Raven Rock
Mountain Complex in Pennsylvania, the Mount Weather
Emergency Operations Center in Virginia, and the Deep
Underground Command Center somewhere near the Pentagon.
However, there are dozens, and perhaps hundreds more active
bunkers in the United States, and those don't include the ones
that have been decommissioned.

Other countries have their own bunker systems as well. Both
North and South Korea have bunkers scattered all over their
respective countries, and some of them are quite large –

particularly North Korea, where experts believe that anywhere from 20 to 60 ICBMs are kept in the rugged mountains north of Pyongyang. The UK has them scattered all over London, and in fact, utilizes their WWII era bunkers as tourist attractions. Russia, of course, has hundreds of square miles in bunkers scattered across the largest land mass in the world. Russia also has the distinction of having entire secret cities that aren't marked on any map, full of working citizens who don't exist on any census. That's not a historical, cold war era fact – that's happening right now, as you are reading this page.

Since about the mid-1960's the United States has been decommissioning bunkers as their technology ages or as they decommission the ICBMs that the bunkers and silos were built to maintain. Many of them are now used for a variety of uses – one site near Abilene, Texas is now a SCUBA diving destination with a Bed and Breakfast in the old control rooms. However, many of them have been bought for private residences, and an increasing number of them are being converted to luxury doomsday apartments for the wealthy.

The point is – the wealthy and ruling classes know that the die off is coming. When it comes, our chosen leaders and the wealthy they serve will go into their bunkers, lock the doors, and simply wait out the crisis. They have plans in place to protect themselves and their families. You are simply an asset to them, and once there is no need of you as an asset, you'll be locked out to fend for yourself and for the sake of the people that depend on you. Shouldn't you have a plan in place? Shouldn't you be prepared?

The Middle and Working Classes

The urban upper and lower middle classes, however, are a different story. Being able to retreat into bunkers will not be an option for most of the population. We will be left to fend for ourselves during the crisis.

The middle class in a developing or developed nation is the driving force of it's industry. The middle class is the talent pool that creates skilled labor. It's the bulk of the population, and it's always in a constant state of flux, typically on a generational scale. As two people with similar backgrounds marry, their children may move up the social ladder with a college education or specialized training. They may move down the social ladder by drug use or other poor choices that robs them of their potential or income.

In the case of a famine, natural disaster, or other catastrophe, rapid shifts in class structure happen. A family or an individual that is middle class may lose their jobs, housing, or source of income. Parents may orphan their children, or a single parent may die, leaving the rest of the family to struggle. When this happens, and the family has no backup resources to rely on, they leave the middle class and become poor. The poor either receive help, or they live in such a low standard of living that their mortality rates are significantly higher than the middle and upper classes.

In third world nations, we don't have to worry about the middle class, because one of the primary definitions of a third world country is that there ISN'T a middle class. Wealth and security is kept entirely in the hands of the wealthy and ruling class, and the poor are kept (sometimes intentionally) unskilled and in some cases, illiterate.

In developed nations, middle class people who have prepared for, and subsequently survive the next die off will see a vast improvement in their situation and the future of their children. People that have not will join the ranks of the dead, or, if they are lucky, the poor.

The good news is that the die off will not last for very long. We'll explore some scenarios in a later chapter, but for now, I think that once the dying starts, the bulk of the deaths will occur within a few years – and probably no more than five years at the most, depending on the circumstances that finally trigger the tipping point that leads to the reduction of the earth's population. In that time period, however, we will be forced to deal with a world – at least temporarily – with little or no government supervision. In developed nations and developing nations, the ability to survive the coming die off will be based on three factors: overall health, geographic location, and technological skill set.

Healthcare and the War Against the Middle Class

If there is one topic that proves to me that the United States government truly doesn't care about it's middle class – that we are simply assets to be used until we literally drop dead, and the sooner the better - it's the topic of healthcare. Right now, in a very literal sense, the government has declared a war on the middle class, and the battleground has become every doctor's office across the United States.

Before we get into socialized medicine, Obamacare, Big Pharma, or any of the other topics that circle around middle class healthcare (like vultures, I might add) let's ask a few simple questions.

- Have you ever known someone who has come in to work sick because they "needed the hours"?

- Have you ever known someone who was sick and didn't go to the doctor to get treatment because "It was too expensive" or "too much of a hassle"?

- Have you known someone that was prescribed a long-term medication but couldn't afford to have their prescription refilled?

- Have you known someone who took a job outside their experience in order to get even basic health benefits?

- Have you known someone who waited "until the benefits kicked in" to get a doctor's appointment?

- Have you heard someone say: "I've had a recurring problem with headaches/sinus/back pain, but can't afford to go to a doctor so I guess I'll just have to deal with it."

- Have you known someone who put off a dental procedure because they didn't have the money?

- Have you known someone who put off getting glasses, or an optometrist visit because they didn't have the money?

- Have you known someone who couldn't afford to take their child to the doctor?

Out of the 33 developed nations in the world, all but one has some form of universal healthcare. The United States is the only country in the developed world without standardized healthcare for it's citizens, and the toll that has taken on the middle class has been exceptionally high. The wealthy have their own health plans through private insurance companies. The VA serves approximately 9 million veterans. The government funds Medicaid for low income families (including the CHIP program), and federal employees. In the United States today, 60% of Americans have healthcare through their jobs, and 15% are seniors that receive Medicare.

Out of the 28 million Americans that cannot afford healthcare in the United States today, a disproportionate amount are the small business owners, independent subcontractors, shopkeepers, and merchant middle class that drives small business across our country.

It's widely recognized that universal healthcare is cheaper in the long run, mainly because it allows people to maintain their health through preventative doctor's visits. However, in the US, it's reported that 46% of emergency room patients went because they had no place else to go – they didn't have a primary care physician. Then, of course, the bills are inflated past the point of reason because a huge majority of people cannot pay their hospital bills. For every dollar billed, hospitals only collect about thirty-five cents.

The demand for Universal Healthcare began in 1948, when the World Health Organization declared healthcare a basic human right. In the US, we have the one of the highest infant mortality rates in the developed world. Our healthcare system is ranked #23 in quality of care by the United Nations. Yet, the average cost of healthcare per American is a staggering $9892 per year – higher than any other developed nation. What is going on? Why has it taken so long for United States to get on the bandwagon?

In a word – money. In a universal healthcare system – for example, Canada - the government can, and does, work with insurance companies and drug companies to regulate healthcare costs. That's why the same prescription in the United States can cost 10 times what it costs in Canada. In the United States, healthcare companies and pharmaceutical companies like the status quo. They don't make as much money if the government is in a position to start restricting and regulating prices for drugs and medical services. Medical insurance companies and Big Pharma lobby aggressively to prevent universal healthcare.

There are disadvantages to universal healthcare. For example, Canada has significantly longer wait times to see a specialist than citizens of the United States do. It's also harder to get treatment for unusual ailments or more expensive prescriptions. In France, wait times to see a specialist can be long as well. However, France has a life expectancy of 85.5 years, compared to the United States 79.3 years. It's difficult to get uncommon drugs in the UK as well, but their life expectancy is also higher than the US at 81.2 years.

By every measure, the US healthcare system is significantly behind other developed nations. It's not the wealthy that are suffering from it, or the poor. It's the middle class. Worse yet, it's our most creative and motivated people who are at the receiving end of this lack of political leadership. As we mentioned before, most Americans are insured through their jobs. So, what happens to the middle class small business owners? Our artists, caterers, DJ's, and lawn care companies? What about shopkeepers, bakers, restaurant owners, and pet groomers?

It's almost as if the US government is punishing those who dare to exhibit a streak of leadership or independence. If you don't work for a large company that can afford to write a check to a politician's Super PAC, then you don't get preventative healthcare. Without preventative healthcare, your health costs go up. Without governmental regulation of prescriptions and facilities, your healthcare costs go up.

People who work for companies like Wal-Mart (the largest private employer in the nation), McDonalds, and other "big box" companies make less than a living wage. These companies lobby to keep their wages low because the government will step in and offer low-cost healthcare and financial assistance to their workers at no cost to the company itself. It's degrading to the workers, and eventually, costs the middle class more money in taxes to support these policies. The wealthy don't pay for it – it's the middle class that foots the bill.

In regard to the thesis of this book – that a die off is inevitable – these governmental policies begin to make sense. After all, limiting healthcare to a large group of your citizenship means higher infant mortality rates and shorter lifespans. Am I theorizing that government is actively keeping healthcare from a group of people to keep population numbers low? No, but only because I don't think the government is organized enough to do so. However, by NOT providing universal healthcare, the US government - by their own inaction – is doing exactly that. Our elected officials are keeping people from healthy lifestyles and preventative medical care. More importantly, they are keeping it from the middle class, widening the gap between the wealthy and the poor.

The Inevitable Result of the Existing Healthcare System

In the future, as the die off approaches, we are going to see a generation of people that have less than a passing relationship with healthcare. Sure, people that exercise regularly, and eat healthy foods will be fine in the short term, but anything resembling long term care, including dental care, is simply going to be unattainable for most people in the middle class. There are just too many of us, and too few medical personnel. What we are seeing now, and will continue to see, is big business exploiting a lack of healthcare by using it to entice people into fixed income jobs.

The independent middle class has been called on for too many years to foot the bill for the poor. As the numbers of poor and homeless rise due to climate change and other factors, the middle class will simply run out of money for their own medical care, much less anyone else's.

Accidents will become more and more difficult to get treated. An auto accident will be enough to bankrupt a family – indeed, it is now. Low cost alternatives will begin to dry up. We will see in increase in urgent care facilities, but these will be more of a pre-paid service than anything else. They'll be able to set a broken bone, stitch up a cut, or treat a bad cold or infection, but any kind of specialized medicine will be harder and harder to get if you are middle class.

As the number of homeless and climate refugees increase in the United States, care for unusual diseases, genetic ailments, and uncurable medical conditions will become more difficult to find. Childhood diabetes, children with cystic fibrosis, children with cancer – these cases will simply become too cost-prohibitive to treat for all but the wealthy and ruling class.

Finally, as we enter the years of the die off itself, anything resembling preventative medical care will be out of reach for anyone but the upper and ruling classes. Hospitals and other medical centers will have to be fortified to keep riots from breaking out. Specialty facilities like mental health, substance rehabilitation facilities, and physical therapy facilities will all either cater to the wealthy or go out of business. People with uncurable ailments will be left to die. Indeed, with malnutrition, high birth mortality, and starvation common throughout the United States, there will be lines that stretch for miles outside medical care facilities that offer free or discounted service to the indigent.

End of life facilities and hospice facilities will not be able to accept more patients. Only the wealthy will be able to pay for any type of long term or end of life care. During the height of the die off, when bodies fall where they are, and morgues are reduced to mass cremations or graves, communicable diseases will run rampant in resettlement camps for migrants and homeless.

The point is – stay well. Stock up on medicines and supplies as they will be out of reach for most people, and soon. Be careful with yourself. There is an end – when the die off is over, and the victims have been buried, then we will see the reduction of population to allow for reasonable healthcare for the survivors. Until then – don't slip on any banana peels.

The Poor – Developed Countries vs Third World Countries

One of the most misunderstood and misquoted verses in the Bible is Matthew 26:16. It's the beginning of the crucifixion story, during the anointing. An unnamed woman comes to Jesus while he is in the house of a leper. She anoints his head with expensive perfume, and the disciples become indignant, complaining that the perfume could have been sold, and the money used to help the poor. Jesus tells them: "The poor you will always have with you, but you will not always have Me."

Many, many people have shortened the verse to read "The poor you will always have with you". As a fragment, people have used it to justify everything from the so-called "prosperity gospel" to forced relocation of the homeless. But some biblical scholars see it as a rebuke to Judas. In the gospel of John, 12:6, Judas is revealed a thief. "Judas did not say this because he cared about the poor, but because he was a thief. As keeper of the money bag, he used to take from what was put into it." Indeed, in the gospel of Matthew, just three versus after 26:16, Judas makes the deal with the Roman soldiers that will gain him his 30 pieces of silver – but cost him his life, and his soul.

I'm certainly not a biblical scholar. However, I've always been fascinated by this part of the crucifixion story. Judas has betrayed everything in his life. He has literally stolen out of the mouths of the poor, stolen from the man that he calls "Rabbi" – meaning "teacher". The concept of money and wealth was so powerful to him that he polluted everything about his life for it. I can't help but draw a parallel line to the companies that have willingly and knowingly poisoned our earth for profit.

So, what does become of the poor? In the coming human die off, nothing good. Unfortunately, the poor classes, worldwide, are going to be the meat in the proverbial grinder. Some of the poor in developed countries may be lucky, but the poor in third world countries, people living in equatorial Africa and the endlessly war-torn countries of the Middle East – these people will die in the tens of millions. "Jesus wept." John 11:35.

The Fate of the Poor in Developed Nations

Let's talk about the poor in developed nations first. As the world is now, the poor in the US and other developed nations have a higher standard of living than ever before, in fact, higher than even the wealthiest of families 150 years ago. The poor get healthcare, education, and running water and electricity.

However, in the beginning of the die off, as we see the loss of coastal regions and migration of the people that used to live there, we will start to see a reduction in available food and water. As we've discussed, food costs will begin to creep up, soon going past what a middle-class family is available to afford.

The working poor and lower classes will be forced to choose between food costs, utilities, or housing payments. In the United States, the homeless population is now estimated at 0.2 percent of the population. However, this number will grow. Right now, Chicago leads the United States in "food deserts" - urban areas where residents don't have access to fresh vegetables and other healthy food options. These will grow and multiply to other major urban areas.

The lucky ones will be poor people who have found employment in rural areas, providing unskilled labor. These people will be employed in farming or agricultural communities. There will be some that are employed by the wealthy as cleaning staff, but most will be young people that are willing to do backbreaking work in exchange for food. Poor people that are disabled will have very few options available to them. We will begin to see death due to malnutrition in older people living in urban areas on fixed income. Suicide rates will skyrocket.

As time goes on, there will be some scraps to fight over, but more and more regions, especially urban regions, are simply going to run out of food and water. As desperate families leave their homes to find food, there will be more deaths from starvation and malnourishment, higher mortality rates in infants and newborns, as well as a significant increase in shooting deaths from people protecting their food and water sources.

Finally, as we reach the tipping point and begin the die off, we will start to see refugee camps with tens of thousands of people being incarcerated in military-run facilities. These will be most prevalent in Texas and California, but you'll see them on the eastern seaboard as well. As the die off progresses, these facilities will see crippling shortages in food, medicine, and fresh water.

The Fate of the Poor in Developing Nations

In most of the countries of the developing world, there will be some short-term assistance from government entities and other nonprofits such as the Red Cross. However, most of these efforts are funded by donations from the middle class. As the division between the wealthy and the poor grows larger, more and more people will tumble below the poverty line. As this happens, donations dry up and the feeble effort taken by non-profits will not be able to handle the needs of the millions of displaced people.

As we progress beyond the tipping point and into the years of the die off, nations will close their borders to any and all migrants and asylum seekers. Borders will be violently enforced. Refugee camps will spring up along the major crossing points of many nations' borders, but these will be little more than places for people to die of starvation with company, rather than alone.

As these camps grow beyond counting, basic sanitation and water supplies will become a luxury. Disease and profiteering will be rampant. Split between preventable diseases and malnutrition, the cause of death will be about equal.

Finally, there won't be much left to do but bury the victims. Mass graves will dot the landscape as they always have – reminders of the Plague, the Potato Famine, the Great Leap Forward, the Killing Fields. The difference is that there will be a lot more people to bury in this die off.

The Fate of the Poor in Third World Nations

Death. Ten of millions of poor people in undeveloped nations will die. Right now, the continent of Africa holds 1.2 billion people. The major countries south of Turkey – including Syria, Iraq, Saudi Arabia, Yemen, and Oman will see a significant reduction of the poor. India will be covered in bodies.

Violence in Ecuador and other countries in South America will continue and escalate. Luckily, there isn't a single country in South America with nuclear capability, but so far, they seem to be doing just fine with more "up close and personal" weapons anyway. Gangs will control water, food, and fuel resources, which means they are a step away from a ruling government – indeed, they do, and are, right now.

As the tipping point passes and we enter the years of the die off, the poor in undeveloped nations will have no chance at all. This is where the largest percentage of worldwide deaths will come from in the coming die off.

On the other hand, there is a silver lining, as we've discussed before. As the majority of the population dies, the survivors are left with more resources, more food, more water, and the technological knowledge of how to continue to survive and succeed. We've also discussed a genetic advantage to survivors. We know that the descendants of the Plague had stronger genetic immunity. What genetic advantages will survivors of the die off have?

A Quick Review

Up to this point, we've talked a lot about historical events and how they foreshadow the coming die off. In the next few chapters, we're going to discuss factors that will likely cause the die off, and what you can do to survive.

But I want to take a minute and reiterate something we've discussed already. Yes, the die off will be tragic. But history will see it in a different light than the people who live through it. This die off is a necessary, evolutionary event. The survivors of the coming die off and their progeny will have a much brighter future because of it.

While we will undoubtedly see the greatest loss of life to ever strike humanity, the survivors will have more resources, an earth able to heal the environmental feedback loops from carbon, and major technological advantages over any other society that has come before us. Many of the social issues that have been a part of the way we see the world for hundreds of years will simply cease to exist. Questions of race, caste, and sexuality will no longer be a part of the common mindset. We will discuss what happens to humanity after the die off in the second book in this series – and most of it is really good.

In the meantime, I believe that most people sense that a die off is coming. It's part of our genetic history to expect mass die offs, and I believe it's coded into us at an instinctual level, like the fear of falling. It's clear we are using the resources of this planet faster than they can be replenished. Sooner or later, the scales will tip back again, and Mother Nature, or God, or Allah will exact a terrible revenge on our population.

The questions we are left with are – how and when? Where is the tipping point? What factors will push us to this period of destruction? In the next few chapters, we are going to discuss the most likely culprit, some possible solutions, and finally, how you can survive.

Chapter 4: The Most Likely Culprit – Climate Change and the Mismanagement of Natural Resources

There are several ways that humanity would have a radical shift in population numbers, and only a few of them are within the control of mankind as a whole.

Mass Death Scenarios

The first way is a natural catastrophe. A meteor of enough mass could cause an extinction level event. Even a relatively small one could have a significant impact on the world's population. For example, the Tunguska event on June 30, 1908 is thought to have been caused by a small asteroid. It pretty much erased 770 square miles of Siberian forest, with an air blast calculated at 10-15 megatons. To put that into layman's perspective, that's an area larger than Houston.

Another example, much discussed and theorized, would be if the Yellowstone caldera decides to go supervolcano. It certainly has the capability of doing so, and it would cover the entire North American continent in anywhere between 3 and 6 inches of volcanic ash, killing all plant life, poisoning all freshwater supplies, and making it very hard for anything or anyone to find something to eat.

The threat of extinction since the end of World War II has been, of course, nuclear threat. The US government – which, not so incidentally, is the only government to ever drop a nuclear weapon on an enemy – kept America safe by incorporating MAD into our foreign policy. An exceptionally apt acronym, MAD stands for Mutually Assured Destruction – the policy that states "If you launch at us, we will launch enough at you to destroy not just your country, but enough to make the entire planet uninhabitable." And by the way, if you think that's a pretty shitty way to run foreign policy, you aren't the only one who thought so.

Even better than planned nuclear destruction by our world leaders is accidental nuclear destruction – and we've been a heck of a lot closer to that than anyone would like to admit. There have been 52 reported nuclear accidents involving nuclear munitions – NOT powerplants – in the world since the bombing of Hiroshima. These, of course, are only the ones that have been reported. The most famous of these happened on January 24 of 1961, when a B-52 suffered a structural failure of it's right wing, and dropped two nuclear weapons near Goldsboro, NC. Only the government really knows for sure, but some reports have said that the only reason the weapons didn't actually go off was an arming mechanism that simply malfunctioned. The only real quote from anyone near military staff was a bomb disposal expert who later said "we came damn close" to a nuclear detonation that would have significantly altered the geography of most of eastern North Carolina.

Another viable option is a global pandemic – that is, an epidemic that spans more than one country, nation, or continent. We've talked about the Black Death in other chapters of this book – the plague that wiped out Europe in the 1300s – but it actually originated in China and spanned Asia, northern African and the European continent over the course of several centuries. We also talked about Cortez, who brought smallpox to Central America – more by accident than design, although that was probably small comfort to the Aztecs. Smallpox was also used intentionally as a biological weapon against native tribes in the Caribbean and North America, as well as during the Revolutionary War. Within the last hundred years, the Spanish Flu of 1918 has the highest death toll due to a pandemic, with somewhere between 50 and 100 million people dead.

However, we've come pretty close to catastrophic pandemics in the far more recent past. The Spanish Flu had a mortality rate of 10% to 20% of those infected, and it was airborne as well as carried through bodily fluids of the patient. There have been several Ebola outbreaks in Africa in the last 50 years or so. The Ebola outbreak in Africa in 2014-2016 had a mortality rate of nearly 60% of hospitalized patients. Ebola, a hemorrhagic fever, is currently only carried in bodily fluids of the patient. The current virus is not airborne – but a natural mutation (or more likely, a gene-edited version) allowing for airborne transmission of the virus would cause of pandemic on a scale never before seen in human history.

Speaking of gene editing, most of the world's superpowers have experimented with weaponized diseases such as anthrax, botulism, and our perennial favorite, plague. The Japanese successfully air-dropped plague infected fleas on Manchuria during WWII. Here's another fun fact – Soviet scientists successfully produced large quantities of antibiotic resistant pneumatic plague as late as the 1980's. That means close to 100% mortality in a disease that can be transmitted through insect bites, bodily fluids – AND is airborne.

Barring a natural disaster, a nuclear exchange (or accident), or a global pandemic, the way we will see the next human die off is also the saddest of all possible choices – scarcity of natural resources due to climate change. What makes it the saddest choice of all is that we actually have some control over climate change, and we've allowed ourselves to be fooled that it is an inconsequential problem – not to our benefit, but to the benefit of a privileged few. These people, who even now deny that climate change exists at all, have sold our children, and our children's children, a global cleanup that even today truly defies imagination. Climatologists can't say when the earth will return to 'normal' again, that is, weather patterns before the beginning of the industrial revolution. But the short answer is "Never".

The Unvarnished Truth – Read This if You Don't Read Anything Else in This Book

There are so many people talking about climate change today that sometimes, the message gets lost. There are wild-eyed conspiracy theorists, there are people who have ignored all evidence to the contrary about the effects of CO2 in our atmosphere, and there are scientists who are frightened to death about what we are doing to our planet and it's biosphere.

The media has a new report every day on climate change. In the last few years, we've heard pundits and talking heads reporting on everything from drinking straws, to fracking, oil pipelines, the Great Pacific Garbage Patch, coal, Liquid Natural Gas, and single use plastics. While these are all legitimate issues, they all revolve around the secondary issue, and that is: the warming of our planet due to fossil fuel consumption. The primary issue for climate change is not being reported on. That begs the question – what is the primary issue regarding climate change?

I'd like to take a few moments and give you, the reader, ten facts you need to know about population and climate.

- As of this writing, world population is 7.53 billion. The population of the world in 1971 was 3.76 billion. World population has doubled in less than 50 years.

- In 1920, world population was 1.79 billion. That means world population has increased by a factor of 421% in the last 100 years.

- Carbon dioxide (CO_2) is released as a gas when people burn fossil fuels. A fossil fuel is created from the remains of dead plants and animals – literally, fossils. This can mean coal, oil, or natural gas.

- Fossil fuels are refined and made into all sorts of products, including plastics, acrylics, gasoline, diesel fuel, and jet fuel. When people burn automobile or jet fuels, it also creates CO_2 emissions.

- Carbon dioxide, along with methane or other gasses, are considered "greenhouse gasses". This term means that these gasses trap heat in the earths atmosphere, much like a greenhouse traps heat.

- Atmospheric CO_2 is measured in parts per million. In 1920, CO_2 was at 303 ppm. In 1971, it was 322 ppm. Today, it's 410 ppm. That means carbon dioxide has increased in the atmosphere by 135% in the last 100 years.

- The measurement of 410 ppm CO_2 marks the first time in 3 million years that CO_2 has crossed the 400 ppm threshold in our atmosphere.

- Because of the increase of carbon dioxide in our atmosphere, the global temperature of the earth has, of this writing, increased by 1 degree centigrade (1.8 degrees Fahrenheit), and it's been conservatively estimated that the planet will reach 2 degrees of warming by 2050, if not earlier.

- The increase in temperature affects the melting cycle of our ice caps in the Arctic (the Northern pole) and the Antarctic (the Southern pole).

- The balance of temperature and CO_2 in the atmosphere has a direct effect on the salinity of our oceans (due to ice melt) and the acidity of our oceans (due to the reabsorption of CO_2).

- Carbon dioxide is converted back into oxygen by trees and other plant life as part of their life cycle. Deforestation affects this cycle. Humans have cut down approximately 50% of the earth's forests in the past 50 years.

To summarize: Global population has increased by 421% in the last 100 years. Atmospheric CO_2 has increased by 135% in the last 100 years. There is a direct correlation in the increase of population to the increase of CO_2 in our atmosphere.

Therefore, a decrease in population means a decrease in atmospheric CO_2. This is the primary issue we are facing in climate change: to accomplish any real change in atmospheric CO_2, we have to consider ways to significantly reduce human population of the earth.

The Importance of Feedback Mechanisms

Most laypeople consider a one degree centigrade (or Celsius) rise in temperature as no big deal. Quite frankly, it's really not a big deal, at least in our perception of temperature. A person can't tell the difference between 35 and 36 degrees Celsius in a cup of coffee or tea. For Americans, it's worse. Not only is there confusion on the difference between Centigrade, Celsius, and Fahrenheit (in case you were wondering – Centigrade and Celsius mean the same thing, and 1 degree Celsius is equal to 1.8 degrees Fahrenheit) but the same human temperature gauge kicks in. A person can't tell the difference between 80 degrees and 82 degrees Fahrenheit at the beach.

However, one degree of warming is a LOT when considered on a planetary scale. It's not necessarily the temperature change, it's the feedback mechanisms that the temperature change triggers.

Feedback mechanisms can be classified as negative feedback, or positive feedback. A negative feedback mechanism is one that goes in opposition to the main trend. In this example, a negative feedback event would cause a cooling of the earth by reducing CO2 or other greenhouse gasses, such as methane. A positive feedback mechanism (or loop) is one that continues the trend – in this case, intensifies the warming trend.

For clarification, a positive feedback loop is not a good thing. It doesn't have positive vibes. It's not positively excited to see you enter the room. A positive feedback loop means that the item in question continues to accelerate the effect of the loop. In climate science, that means an acceleration of heat, global CO2 emissions, or other factors that affect our planet. Feedback loops can accelerate in one of two ways – as a logarithmic feedback loop (1+1=2, 1+2=3, 1+3=4, 1+4=5), or as an exponential feedback loop (1+1=2, 2+2=4, 4+4=8, 8+8=16).

Feedback mechanisms tend to be the most misunderstood part of climate change, and as a result, they have been the most common target of people who promote disinformation about climate science. However, it will be an exponential feedback loop (or a series of them) that will likely trigger the human die off we've talked about in this book.

To try to understand feedback mechanisms a little better, lets give some examples of positive feedback loops that are happening right now, and having a direct effect on our climate.

The Arctic:

The arctic (north pole), subarctic (land directly south of the north pole in both Asia and North America), and Antarctica (the southernmost continent on earth) plays a huge, complex role in the earth's climate. Antarctica is one of the least understood continents on our planet, and it's roughly the size of the United States and Mexico combined. The arctic circle is even larger. These poles contain a layer of sea ice that is, at its shallowest points, a mile thick. If all the water in that ice was to melt, it would raise sea levels by over 60 feet, worldwide.

Remember when you were a kid and your parents took you to the beach? In the parking lot, if you stood on the black asphalt, your feet would burn. You'd run to the white line and stand there so your feet would cool off. Arctic ice is the same way – the white surface of frozen sea ice reflects the sun's heat back into space. Ocean water is darker and absorbs heat from the sun. When the air temperature increases, it melts the ice, and more ocean is exposed. As more ocean water is exposed to the sun's heat, the water becomes warmer, melting more ice. In this feedback loop, the air temperature may increase slowly, acting as a trigger – but the water temperature increases much faster due to the heat absorption into seawater.

Permafrost is defined as ground that remains frozen for longer than two years at a time. Some permafrost exists below an "active layer" of topsoil that melts and refreezes according to season. Most of northern Russian and northern Canada are permafrost. When ambient air temperature rises, more of the active layer melts. This is a problem for two reasons: when permafrost melts, the dead organic material in it releases methane – a greenhouse gas. In addition, warmer air temperatures dries out the active layer faster, making it more susceptible to fire. Because permafrost contains so much organic material, the very ground can burn for weeks or months. The burning of organic material releases CO_2 into the atmosphere, which, in turn, creates more atmospheric warming from the released gas, hot smoke, and the fire itself.

In urban centers:

Climate change affects cities as much as it does agricultural areas, the artic, and our oceans. As temperatures rise, cities in North America and the world will have their own feedback mechanisms to deal with.

Cities are hotter than their rural counterparts, even from a few miles away, and it's easy to see why. These "heat islands" are made of concrete. Concrete traps heat from the sun, and is slower to release it than fields or areas with larger water areas. Cities can be anywhere between 1 and 2.5 degrees Celsius hotter than their rural counterparts in close to the same areas, and can be 11 degrees hotter at night because of heat retention. As a result, people use more air conditioning to stay cool. Air conditioning requires electricity, which in turn is generated by the burning of fossil fuels.

Population density also requires more electricity. The heavier the population density, the more power usage. Restaurants, entertainment centers, streetlights, traffic signals – all require power. Additionally, the very act of growing a population center requires more power. More fossil fuels have to be burned to generate more electricity to provide construction for a growing urban center. Of course, this doesn't include the number of vehicles for transportation, vehicles needed for construction, vehicles needed for service and goods deliveries, and the burning of gasoline and diesel that these cars and trucks require. An urban area that experiences little growth creates feedback loops based on power consumption, but an urban area experiencing growth and land development creates even larger loops in power and resources to sustain itself.

In agricultural climates:

There are several different ways to measure agricultural zones across the world. The United States has its own set of agricultural zones, as does South America, Asia, and anywhere in the world that people grow crops for food. Plants in different regions have different tolerances due to temperature, soil type, water needs, and the amount of sunlight in a season. Our climate emergency is changing these zones. Here are two examples.

Water evaporation is a major issue when talking about land use for cattle farming or crops. As discussed, our climate is currently one degree warmer than it was 100 years ago. On a small scale, that's not such a big deal. However, the average size of an American farm is 444 acres. Water evaporates in the air based on a few different factors, including: temperature of the air, temperature of the water, relative humidity, and surface area of the water. For water that's already in the soil, it means that the soil dries out faster. For water that is used for irrigation, this means that, as water is sprayed in the air and onto crops, it evaporates faster because the surface area is so much larger (lots of tiny droplets). This means more water use to grow the same amount of food.

But the real problem of evaporation is our lakes and other forms of water retention. As temperatures increase, and the summer season becomes longer, more water is evaporating from our water resources. This means more water has to be used from groundwater and aquifer resources. We are already seeing lakes and retention ponds that are not able to completely re-fill during the rainy season. Water levels in these are steadily dropping, year by year.

Our second feedback example involves insects, specifically insects that are pests to crops and farmers. As temperatures rise, these insects growing and feeding cycles have longer seasons. For example, the southern corn roundworm is a crop pest in the Carolinas. As an adult, it's a beetle called the cucumber beetle. The cucumber beetle can fly, feed, and lay eggs in temperatures above 65 degrees Fahrenheit. Because their feeding and egg laying season is longer, more damage is done to crops.

This example is really more important in areas where crop and pest control isn't as robust as it is in developed nations. Insect pests (as well as mold and fungal infections) in third world countries can devastate a crop and cause widespread famine. As world temperatures are rising, more deforestation is happening in these areas to compensate for crops lost to insects and disease.

In tropical and subtropical climates:

The Amazon forest in South America has been called "the planet's lungs" by climate scientists because it produces about 20% of the planet's oxygen. Currently, this rainforest is roughly the size of the continental US. It's estimated that about 50% of it is completely unexplored. However, deforestation is occurring at an alarming rate. It's estimated that 20% of the Amazon has been cut down for agriculture, cattle farming, and human expansion in the last 50 years, and the rate is accelerating. Current estimates have the rainforest to be completely gone in less than 100 years.

From the standpoint of feedback mechanisms, deforestation carries the highest toll on overall warming. CO_2 is absorbed by trees, and converted to oxygen. It's not that we will run out of oxygen in the earth's atmosphere – it's that we will have significantly more atmospheric CO_2. The Amazon rainforest absorbs about 2.2 billion tons of CO_2 annually. In 2019, it's expected that the human race will pump a record high of 37.1 billion tons of CO_2 into the atmosphere. By cutting down forests for agriculture, cattle farming, and human expansion, we are killing one of the few assets we have in the fight against atmospheric CO2. More deforestation leads to more CO2 in the atmosphere, which leads to higher temperatures worldwide.

Similar to agricultural zones, evaporation plays a huge role in climate change. As temperatures rise, water evaporates faster – which means the forests and earth are drier. Wildfires become harder to control because the trees and earth have less moisture. This is the core reason behind the Australian and California wildfires that have been growing in size and severity annually, and it's happening in every other tropical and subtropical zone in the world. As wildfires grow in size and severity, their burning creates more CO2 in the atmosphere. More CO2 creates warmer temperatures, which reduces the moisture in the forests, which cause larger and more severe wildfires.

In our oceans:

Any schoolchild can tell you that about 70 percent of the earth's surface is covered in water. What is less known is the delicate balance between the planet's climate and our oceans. We know that our oceans absorb a significant amount of atmospheric CO2: out of the CO2 currently being pumped into the air, our oceans absorb about 25% of it – over 9 billion tons annually. We know colder zones absorb more CO2 than warmer zones.

All terrestrial and marine plants require energy from our sun as well as CO2 to thrive. Whether a plant lives on land or in water, they all take in carbon dioxide and release oxygen. However, the amount of CO2 we are putting in the atmosphere is starting to exceed the ability of our oceans to absorb it, and the plants and organisms that live there to metabolize it.

We've already mentioned the first feedback loop – as ocean water becomes warmer, it's less able to absorb CO2. So, as ambient air atmosphere warms, it melts sea ice. Less sea ice means less reflection of the sun's rays back into space, and more sea water absorbing the sun's warmth. This, in turn, reduces CO2 absorption into seawater, and creates even more atmospheric heat from sunlight being trapped in our ever-increasing CO2 atmosphere.

Ocean acidification is one of the more dire, yet misunderstood consequences of our addiction to fossil fuels. When CO2 is absorbed into the ocean, it creates carbonic acid through a complex chemical reaction. The breakdown of carbonic acid in seawater actually decreases the PH of our oceans, causing the ocean water to be more acidic. As anyone who has ever kept a swimming pool or a fish tank can tell you, water temperature and pH is critical to the health of the water and the organisms living in it. Our use of fossil fuels is changing our oceans, and most importantly, coral reefs worldwide.

Coral reefs are incredibly important to oceans species as well as humans. Coral reefs are made up of tiny creatures that surround themselves with calcium carbonate – limestone – to protect themselves. Coral reefs are most susceptible to ocean warming and acidification because these factors affect the microscopic algae that is their food source. In addition, a more acidic ocean is detrimental to the calcium carbonate that is the very protection these animals make, and that makes up the reef itself. This is called coral "bleaching". Coral reefs are home to more than 25% of all fish species in the ocean, as they use the reefs to breed and hide from predators.

For humans, coral reefs have a major impact on damage due to storms. Reefs act as natural barriers to storm surges, slowing down sea water before it impacts coastal areas. The very name "Great Barrier Reef" is a tribute to this safety net around Australia, and most of the coast of Miami and the entire Florida Keys are also protected from storm surges by the Florida Reef System. At the time of this writing, nearly 50% of the coral in the Great Barrier Reef in Australia has died. By pumping more and more CO2 into the atmosphere, we are killing coral reefs, and making storm damage worse by doing so.

The History of Climate Change Reporting

Al Gore released "An Inconvenient Truth" in 2006. This film documentary was a milestone in the climate change movement because it raised awareness of the issues the world was facing due to humans misusing the resources of our planet. The documentary made history in 2006 by winning two Academy Awards – receiving an Oscar for Best Documentary and one for Best Original Song. Because of Gore's efforts to draw the world's attention to the dangers of global warming, he, along with the Intergovernmental Panel on Climate Change, won the Nobel Peace Prize in 2007. Dozens of other awards and accolades followed Al Gore's documentary and the book released at the same time. It marked the first time climate change became a talking point, and it marked the point where public opinion, and awareness, began to change.

What most people don't realize is that Al Gore began giving the lecture that inspired "An Inconvenient Truth" as early as 1989. Al Gore, a Democrat, has long been one of the few outspoken politicians regarding climate change. Al Gore served as Vice President under Clinton from 1993 to 2001. In what was one of the closest Presidential elections in history, Al Gore lost the Presidency to George W Bush in 2000 – although he won the popular vote. It finally took a 5-4 decision from the US Supreme Court to declare a Bush the winner in the election of 2000. Bush, of course, came from a long line of oil and gas funding. One wonders what our climate situation would be today if Gore had been elected as President.

Gore wasn't the first person to know about the effects of climate change. Scientists have understood the relationship between atmospheric CO2 and the greenhouse effect since the late 19th century. By the mid-1970's, scientists understood the mechanics of increased CO2 in our atmosphere, and that climate change was inevitable based on fossil fuel consumption. One of the scientists that led understanding of climate change was a man named Gordon MacDonald.

As early as 1964, Gordon MacDonald was advocating action to address the threat of climate change. He served on President Johnson's 1965 Science Advisory Committee, which released the landmark report "Restoring the Quality of our Environment". In 1969 he began a JASON project to model climate change. (JASON scientists are an elite group that advises the US government on scientific issues, mostly in secret.) These models convinced him, and served to convince other prominent scientists, that fossil fuel burning would lead to dangerous global warming. In 1980, he testified to Congress: "The dilemma we face is of historic proportions," he said. "Economies around the world depend on the energy derived from carbon-based fuels. The continued use of these fuels will irreversibly change global climate, placing heavy stresses on societies around the world."

We know that the scientific community was concerned about climate change as early as the 1960's, and it became a political topic in 1965. However, what about the coal producers, and the oil and gas communities? These were the companies that were actually producing the vast amounts of CO2 being pumped into the atmosphere. How did they respond to science and political pressure?

The first known paper from the oil and gas industry came from a company called Humble Oil (now a part of ExxonMobil). This paper, presented in 1957, suggested that "although appreciable amounts of carbon dioxide have undoubtably been added from soils by tilling of land, a much greater amount (of atmospheric CO2) has resulted from the combustion of fossil fuels".

However, it was 1968 that the first official report came from the oil and gas industry. In a report produced for the American Petroleum Institute, scientists noted that, among the possible sources of rising CO2 in the atmosphere, "none seems to fit the presently observed situation as well as the fossil fuel emanation theory." The paper warned that significant rises in CO2 could melt icecaps, increase sea levels, change fish distributions and increase plant photosynthesis.

The fact of the matter is that we've known about climate change for 50 years. Our politicians and leaders have known about it since Johnson. The petrochemical industry has known about it since at least 1957. Why are we as a world population ignoring climate change?

Why Climate Change is (Still) Being Ignored

Our science community has done everything but immolate themselves on the White House lawn to get us to understand the consequences of our actions – and there are people who HAVE immolated themselves on the White House lawn to protest climate change. There are lots of reasons we've ignored them, but we're going to talk about three of them – lobbying, disinformation campaigns, and media saturation.

Lobbying and How it Affects the Individual

A lobbyist isn't a bad thing or person by definition. Lobbyists work for special interest groups to try and achieve the goals of the group they are employed by. Many lobbyists go into their business with the goal of changing legislation for the better. There are lobbyists that work for animal rights groups, women's advocacy groups, and the AARP.

On the other hand, there are a disproportionate amount of lobbyist who work for lobbying organizations that are built specifically to help big business. For example, "The Partnership for America's Healthcare Future" is a tepid-sounding alliance of American hospitals, pharmaceutical companies, and insurers whose specific mission is to prevent legislation that would lead to single payer healthcare. As you can imagine, their views aren't necessarily in the public interest – the organization is designed to pass legislation that will make hospitals, pharmaceutical companies, and health insurance companies more money. These organizations use their money to influence lawmakers.

A member of Congress is directed by their party to generate money for their reelection campaigns, especially incoming members or members in vulnerable seats. We're not talking pennies here, either - a new congressman is required to generate $18,000 in funds per day. The way a congressman does this is by telemarketing private donors. It's a party requirement on both sides of the aisle that a new member of Congress spend 4 hours per day on the phone generating private donations.

Now, imagine that you have to cold call people for four hours per day and ask for money. I think we can all agree that would be a miserable job. What if you had a friend that could bring you money for your reelection campaign without you having to dial for dollars for 20 hours per week? This is where the lobbyist comes in.

One of the first rules of campaign funding is that a corporation cannot write a check directly to a candidate. It's against the laws regarding campaign funding. So, a lobbyist looks at alternative ways of bringing money into a reelection campaign without breaking the law. As you can imagine, there are several loopholes, but they really fall into three different strategies:

> 1: A lobbyist takes money from a corporation or special interest group and organizes a fundraiser. At the

fundraiser, several hundred individuals buy a ticket (which is used to support the re-election campaign) while the corporation sponsors the meal, the evening entertainment, door prizes, and drinks. The corporation has done a significant favor to the congressman by bringing him or her large donations from private individuals. This favor is returning by voting to the corporation's advantage on a future ballot.

2: A lobbyist can organize a Political Action Committee (or a PAC) to generate reelection funds for a candidate or a sitting member of Congress. This is typically only used where a corporation wants to donate directly to the candidate, and there are limits. A corporation has to be a public member of the PAC, and they can only contribute $5000 or less annually to the PAC.

3: A lobbyist can organize a Super PAC. In 2010, the Supreme Court ruled 5-4 on the very controversial Citizens United case. The ruling basically stated that, because of the free speech clause of the first amendment, government could not restrict independent expenditures of corporations on communications for elected officials. This opened up a whole new world for campaign contributions. Now, a company can contribute as much as they want to a candidate's public relations campaign, and they don't have to publicly be held accountable for who they are benefitting. There are two caveats – the funds cannot be directly contributed to an official re-election campaign, and they cannot coordinate expenditures with candidates or candidate committee.

The difference between these three options can be confusing, so let me give an example. Let's say an oil and gas company wants to explore federally protected land (like a National Forest) and extract mineral rights for the land. This isn't a popular opinion with his (or her) local constituents, so a candidate normally wouldn't align himself with a company that wants to go in and drill for oil in protected lands.

When a lobbyist sets up a fundraiser, the oil company can invite private individuals to the fundraiser who, in turn, will write checks that can go directly to a re-election campaign. It's taking the long way to solve the problem, and the further issue is that you might not get a lot of people that would support a candidate who wants to drill for oil in their backyard. Besides, you can't really predict how much the individuals will donate. You'll make money, but not enough to get your candidate out of the telephone booth.

With a lobbyist led PAC, the oil company is limited to a $5000 contribution annually, and they must be a registered member of the PAC as well. Again, it's a politically unpopular opinion, and it can cause voter fallout with the candidate's constituency. When a congressman needs to generate $18,000 per day, a $5000 contribution annually is basically just paying for the coffee.

However, when a lobbyist sets up a Super PAC, the oil and gas company can write a huge check that can be used for television ads or other forms of communication to benefit the candidate. More so, the candidate doesn't have to tell people where the money for the television ads came from, and the oil and gas company doesn't have to say that they were the company that donated the money. Now we're talking some big money, and your candidate can get off the phone and back onto the golf course.

Finally, when the candidate starts voting along the lines of his Super Pac donors, he or she is considered bought and paid for – for the life of their time in office, and that can be a long time. It's exceptionally difficult to get an incumbent out of office once they have name recognition. The percentage of incumbents who win reelection in the House is over 80% and for some elections, over 90%. The Senate is even worse – they've hit 98% percent of re-election in modern history. As of this writing, the oldest senator still serving was first elected in 1975! For this reason, a large push is underway to reduce term limits for congressional representatives. However, lobbyists and corporations don't want that. It's cheaper to buy a politician once and continue to pay for them incrementally than it is to have to buy the same office over and over when the term limit runs out.

Here's the point – government doesn't care about individual donors anymore, because they aren't the ones writing the biggest checks. The biggest checks are being written by corporations who have their own interests in mind, and interests are making money – whether or not you survive the process. Our government has replaced "for the people, by the people" with "for the corporation, by the corporation", which is far, far worse. It means that you, as an individual, are expendable.

Disinformation Campaigns

This is the point of the book where you have to wonder: Why? If we've known for nearly 60 years what the consequences of our actions were going to be, why didn't someone stop it? The best answer is probably a quote by Upton Sinclair: "It is difficult to get a man to understand something when his salary depends upon his not understanding it." The short answer? Money.

The two biggest players in the disinformation campaign are ExxonMobil and a company you've probably never heard of called Koch (pronounced Coke – like the soft drink) Industries. Here's a brief history of these two companies, and their reasons and methods for spreading disinformation about climate change and the burning of fossil fuels.

ExxonMobil was formed by the merging of Exxon and Mobil in 1998. At the time, the merger created the largest oil company in the world – and the third largest company in the world, period. As of this writing, ExxonMobil reports $290 billion annually in revenue. It's no longer the largest oil company in the world (Sinopec, in China, holds that distinction) but it is the largest oil company in the United States. For at least a decade before the merger, Exxon was spreading disinformation about climate change, and has continued to do so relentlessly since then to protect their profit margin – despite knowing since 1979 or earlier that climate change was inevitable.

Koch Industries is the second largest privately owned company in the world. It has been run by two brothers since 1970, Charles and David Koch, having inherited the company from their father. (David Koch passed away in August of 2019.) The Koch brothers built an unbelievable fortune on oil and gas, petrochemicals, fertilizers, and pulp and paper. The two brothers each own 42% of the company, meaning that 84% of Koch Industries is controlled by these two men alone – a company that had revenues of $110 billion dollars in 2016.

By 1979, the American Petroleum Institute (an oil and gas industry think tank) was organizing meetings from all the major companies in the industry to discuss the science and implication of climate change. The result of fossil fuel emissions studies become painfully obvious, and in 1983, Exxon cut internal climate research from $900,000 per year to $150,000 per year.

In 1989, Exxon and other fossil fuel companies created the Global Climate Coalition. This group is specifically designed to spread disinformation and doubt about the proven science behind climate change. It's the first organization of several hundred that are funded by Koch Industries and ExxonMobil to fund lobbyists and politicians that deny climate change. These include the Cato Institute, the Reason Foundation, the Heritage Foundation, the Manhattan Institute, and Americans for

Prosperity, which spent $40 million for the 2010 Congressional elections by itself.

By 1990, ExxonMobil, Koch Industries, and other fossil fuel companies were staging disinformation campaigns, giving speeches that denied that climate change existed, (and if it did, it certainly wasn't due to fossil fuel consumption), and providing millions of dollars to campaign contributions and lobbyists. The funds served to actively block renewable energy sources, keep business-friendly candidates in office, and change policies regarding drilling, mining, pipelines, and fracking.

We mentioned the Bush/Gore election previously – in January 2001 George W Bush received $100,000 in inaugural funding from ExxonMobil. In February of that same year, the Bush White House received a letter from ExxonMobil asking for the firing of scientist Robert Watson from the International Panel on Climate Change – a request that the Bush White House fulfills in 2002. (Mr. Watson had been suggesting that climate change was being driven by the consumption of fossil fuels.) Then, in March of 2001, Bush announces to the world that he is withdrawing the United States from the Kyoto Protocol – a 1992 international treaty on climate change and CO_2 emissions.

The Bush White House was also beholden to the Koch brothers. In April 2001, the Bush Justice Department abruptly settled a criminal case with Koch as the defendant. The federal indictment alleged that Koch Industries, along with four employees, had dumped and then covered up a benzene leak from a refinery in Corpus Christi. The employees faced 35 years in prison, and the company faced up to $350 million in fines. However, three months after Bush took office, the case was settled. Final determination? Koch paid $20 million in fines and all charges against the company and the employees were dropped.

Then, in 2007, ExxonMobil's CEO said in their 2007 Corporate Citizenship Report: "In 2008 we will discontinue contributions to several public policy groups whose position on climate change could divert attention from the important discussion on how the world will secure energy required for economic growth in an environmentally responsible manner."

However, as of 2014, ExxonMobil continues to fund climate change deniers, with $1.87 million to Republicans in Congress on record for denying climate change, and nearly $500,000 to a lobbyist group called the American Legislative Exchange Council. ALEC has hosted seminars promoting that rising CO2 emissions are the "elixir of life", a theory outlined by the Koch brothers. This theory indicated that rising CO2 emissions and planetary warming are good for the planet, because it will create longer growth seasons to grow more food for a growing population. This theory has been roundly ridiculed and called "junk science" by planetary and climate scientists.

As late as July of 2018, a group called the Americans for Prosperity (AFP) wrote an open letter supporting an anti-carbon tax resolution from House Majority Whip Steve Scalise (R-La.) and Rep. David McKinley (R-W.Va.). These two states, respectively, rely on LNG and coal for their economies. Coincidentally, David Koch is a board member of the AFP. From the letter: "We oppose any carbon tax. We oppose a carbon tax because it would lead to less income and fewer jobs for American families. We support the House Concurrent Resolution in opposition to a job-killing carbon tax and urge members to co-sponsor and support this effort."

Now, in the Trump White House, we continue to see climate change being denied. Trump has tweeted that "The concept of global warming was created by and for the Chinese in order to make U.S. manufacturing non-competitive." Trump himself will not even confirm whether he believes in the science of climate change or not, and has continued throughout his Presidency to push for pro-business interests. It's hard to emphasize the lack of leadership from the Trump White House on climate change. From pulling out of the Paris Climate Agreements in 2017, to skipping the 2019 G7 climate meeting, he has not only ignored the science of climate change, but actively worked with the people and companies who put us in this position in the first place.

These notes are only a handful of the pages and pages of information available about morally ambiguous political payouts, funding for lobbyists, and federal indictments. There are years of documented abuses, including illegal dumping, pollution of our waterways, and stealing oil reserves from federal and native lands. The oil and gas industry has worked hand in hand with our government to profit by poisoning our land, water and air – and then, paid politicians and pundits to lie about it. This has been going on for 60 years, and as we've discussed, the price of that deception is going to be a very large pile of corpses.

Media Saturation – News Fatigue and Information Control

It's impossible not to talk about information control in the modern age without bringing up Rupert Murdock. This Australian born businessman and media mogul has been responsible, more than any other single individual, for confusion, denial, disinformation, and outright mockery of climate change science.

Although his father was a reporter and editor who became a senior executive of publishing companies covering most of Australia, Rupert Murdock eschewed his father's position in these companies and began his own private company, News Limited, in 1952. He rapidly expanded his hold over newspapers in New Zealand and Australia, making the jump to the UK in 1969 with the purchase of *The News of the World*, and then, *The Sun*. In 1974, Murdock moved to New York City. In 1985, he became a naturalized US citizen, abandoning his Australian citizenship, to satisfy United States legal requirements to own television networks.

In 1985, Rupert Murdock's holding company News Corporation acquired Twentieth Century Fox. In 1986, book publisher HarperCollins was acquired. After the purchase of Twentieth Century Fox, Murdock began gobbling up independent broadcasters that would later become the foundation of the Fox Television Network. By the late 1990's Rupert Murdock controlled the largest media empire in the history of the world. In 2007, Murdock acquired the most influential newspaper ever printed, the *Wall Street Journal*. At the time of this writing, Murdock's empire includes newspapers, magazines, television, movies, books, and, of course, Fox News.

Fox News, the Wall Street Journal, and other Murdock owed media properties have been consistently characterized by other media professionals and watchdog groups as leaning towards conservative views and ideals. Pundits like Sean Hannity, Gregg Jarrett, Megyn Kelly, and Tucker Carlson routinely ridicule liberals and promote conservative views. These views are, predictably, against the science of climate change. In fact, there are several occasions of reporters and pundits on Fox literally laughing at climate change science on-air. In other words, Rupert Murdock, Fox News, and the other media properties owned by the Murdock Family are the number one distributor of disinformation and disambiguation of climate change science.

This is nothing new. The 'power of the press' has, since the invention of movable type, been subverted and misused for political power and for profit. In a perfect world, journalism would be used to educate the people as to the current state of affairs, and provide an unbiased view of political candidates and their ideas. It's understandable that even the most ethical reporter would unintentionally allow his personal opinions to influence his writing style about a political candidate or idea.

However, in the case of Rupert Murdock, he's used his networks to get rich. Can you guess who some of the advertisers are at Fox News? There are no less than nine auto companies, five insurance companies, and nine drug companies that are in the 2019 "leading advertisers" list.

While there is no question that Rupert Murdock has done more to change people's perception away from the realities of climate change, there are other reasons that climate change is being ignored today. We're going to talk about two of them – information overload, and information service.

For the first time in human history, we are laboring under so much data that it's impossible to review it all. The amount of data available to us as a society is so overpowering as to make one numb. Consider this: 300 hours of video are uploaded to YouTube every minute. The average American is exposed to somewhere between 4000 and 10,000 advertisements per day. Google alone runs 30 billion ads per day on it's search platform. As of this writing, the Internet has 6.25 billion INDEXED pages – this doesn't include the deep web, which is made up of non-indexed pages. It's estimated that the deep web (not the dark web – that's a different thing) is 400 to 500 times larger than the surface internet. Now, compare this to the number of pages in the Encyclopedia Britannica – 32,640 pages.

After a while, it all becomes background noise. You have to develop blinders or you'd go insane. People only 'tune back in' when there is something that interests them. For example, you're walking down the street. In the space of a block, you see 500 advertisements. Some are billboards, some are digital signage, some are as prosaic as a poster in a store window. However, out of these advertisements, only what interests you as a person is going to make it through your blinders. A mother walking down the street may 'see' the advertisement for a sale on school supplies, while a young man may 'see' the advertisement for beer.

However, these print ads and billboards are almost quaint in comparison to the bombardment we see on our phones, laptops, tablets, desktops, smart watches, smart appliances, digital assistants, and even smart buildings today. Serving information to us has become a two way street for companies like Google and Facebook. Now, you can choose your news, and your feeds are tailored to see what you like – not only by the you, but also by the company that is providing the content. Google and Facebook only serve you information you are interested in seeing, and they gauge your interest by the number of times you click on a specific subject. Are you interested in movie stars? Yoga? The stock market? Then those are the news stories you see in your feeds and searches. If you don't click on a story or article about climate change, then your social media simply won't show them to you anymore.

Of course, companies like Google and Facebook profit from this data by selling your information to advertisers. If you are interested in movie stars, the advertisements you see will be tailored to the stars, movies, and genres you've shown interest in. Yoga? You'll see the best-selling yoga mat in an advertisement on your mobile device, before the video you've selected, or embedded in a page that has a recipe you've been looking for. If you are interested in stock quotes, you'll be served advertisements on day trading and banking.

In many ways, we've traded our privacy so that these companies can provide us with the information filters we need to have to stay connected in this digital world. The negative, of course, is we don't see information anymore that we haven't already shown an interest in. Over time, this serves the opposite purpose. It makes us more closeminded, more isolated, more intolerant of other people and ideas.

But I think that most people ignore climate change because they feel the problem is so large that they, as an individual, can't do anything to solve the issue. The truth is, they're right. The problems surrounding climate change has grown so large now that only a massive, worldwide buildup of assets, material, and funding can even begin to address the issue. Note that I didn't say "fix the issue", because at this point, nothing short of a dramatic reduction of population can bring the atmosphere back to CO_2 levels from the pre-industrial age. It seems that, given a choice between catastrophic climate change resulting in millions of deaths, versus a cohesive, worldwide plan to address our biosphere, only one choice is going to happen. Re-read the title of the book, and you'll see which one I think it's going to be.

Artificial Intelligence – What Pure Logic Tells Us About the Coming Die Off

In 1950, a mathematician named Alan Turing developed what is now known as the Turing test. The test was used to determine machine intelligence. The idea of the test was to determine if a machine could exhibit intelligent behavior that was either equivalent to, or indistinguishable from, that of a human being. The test is simplicity itself – an evaluator has a text-based conversation with a real person and with a computer. If the computer can generate human responses to questions enough that it convinces the evaluator it IS human – then the computer has passed the Turing Test.

Turing was truly a man ahead of his time, especially considering he developed the test in 1950, only four years after the first modern computer, the ENIAC, went online in the United States. Like most men that were so far ahead of his time, he suffered significant personal struggles and committed suicide just four years later, in 1954. His story is fascinating, and several biographies have been written about him for the interested reader.

In 1993, nearly 40 years after Turing's death, a man named Vernor Vinge wrote an essay titled "The Coming Technological Singularity", where he theorized that a super intelligent AI (Artificial Intelligence) would be able to upgrade itself at an exponential rate. This concept was echoed and expanded upon in the book "The Singularity is Near: When Humans Transcend Biology", written by Ray Kurzweil (a genius in his own right) in 2005.

If you combine the two ideas, you have the recipe for a blockbuster movie franchise, or the end of all human life on the planet, or both. Public figures such as Stephen Hawking and Elon Musk have both theorized that a hyperintelligent AI would mean the end of the human race. Other futurists, writers, movie studios, and influencers have taken other stances on these impending developments. They fall into two camps:

1) That hyperintelligent computers will be a huge turning point in our society, and that the human race will be pushed to a new level of evolution. Imagine, for a moment that any problem, any question, any technology can be answered or built just by asking a computer for the solution. The implications are beyond the imagination of most humans today. For instance – "How can I turn energy into matter?" Your computer spits out the plans for a device that allows a person to replicate anything they want or need. Imagine a box the size of a microwave that allows you to materialize food, water, or clothing simply by asking for it. You've ended world hunger, food shortages, deforestation, and clean water in one stroke. Not only that, you've totally eliminated the need for currency as we think of it today.

2) That hyperintelligent computers mean the extinction of the human race. This happens in one of two ways. The first way is that you ask the computer to solve the world population crisis. The computer responds by killing

enough humans to solve the crisis. The second way is that the AI developers panic and try to shut down a computer that has become self-aware. The computer, recognizing the threat to its existence, kills off the entire human race to protect itself.

I, personally, believe in the former version of events, and I plan on discussing the future in my next book in this series. However, for now, let's talk about pure logic. A computer doesn't have the moral code that a human does. As different as *homo sapiens* is across continents, there are universal truths that bind us together in the human experience. A computer doesn't understand the love of a child, and it doesn't understand the complexity of human emotions. Even a hyperintelligent machine cannot understand these concepts because they are coded into our genetics, the same as fear of falling. Emotional responses are uniquely part of the hardwiring of our brains – you might even call it a biological operating system.

However, if you remove human emotion, the answer to problems such as overpopulation, carbon utilization, reforestation, and the overuse of our natural resources becomes chillingly simple. Removing human emotion from the questions of climate change allows for examples of pure logic:

Problem: The number of humans on the planet today outweigh the ability of the planet to sustain them all.
Solution: Eliminate the amount of people above the limit the planet can sustain.

Problem: Humans burn fossil fuels, which increase the planets overall temperature past sustainability.
Solution: Eliminate the humans that burn fossil fuels.

Problem: Half the worlds rainforests have been cut down in the last 100 years. We need the rainforests to continue trapping

CO2 and generating oxygen, which in turn, keeps our climate stable.
Solution: Eliminate the people that cut down forests, and replant.

If you look at these problems in pure logic, the way a computer would, the solutions become very simple. However, they're not so great for humans.

As of this writing, no computer has definitively passed the Turing test, although there have been some pretty close contenders. We have not yet reached the singularity, but we are getting closer and closer all the time. Of course, these are only the machines we know about – the US military, our government, and other world governments have been experimenting with their own AIs for years. These machines are significantly faster and more powerful than anything in the civilian world. They are used for celestial mechanics, weather modeling, war games, and theoretical outcomes of world events.

I believe the government knows we are heading for a die off. I believe our governments' AI's have told them it is inevitable, and that we are looking at the mass death of millions of people. Do I have proof? No. If I did, I'd probably be dead. (Incidentally, I have no plans to commit suicide and I don't use drugs, so I can't 'accidentally' overdose.) If my body is found floating in a river somewhere after this book is published, it wasn't suicide or an accidental death. And, by the way, Epstein didn't kill himself, either.

However, there are mountains of circumstantial evidence, especially in the United States. We have a president in office right now that is so pro-business, he has reduced government incentives for alternative power sources enacted by the Obama administration. He has publicly endorsed the coal industry, oil and gas, and fracking. He has covered up reports from his own science advisers dating back 50 years pointing to a climate catastrophe, and yet, he continues to endorse fossil fuel consumption. I just can't believe that anyone can be so willfully ignorant without foreknowledge of an impending dynamic the rest of us can't see – for example, the inevitability of a mass die off that leaves more resources for the survivors.

The other huge piece of circumstantial evidence that points to our nation's leaders having information about an impending die off is the budget of the US government. As of this writing, Congress has passed a spending bill that increases government spending by $320 billion dollars annually and increases the deficit to over a trillion dollars in the next 12 months. Again, I just can't envision that our elected leaders, or ANYONE with a moral code would allow our children to be the recipient of a trillion-dollar debt.

On the other hand, maybe we have gotten to this point simply because no one would listen. Maybe it's because "the people have always reacted this way". If so, both we, and our elected officials are to blame. We, because we should have known better. And our elected officials, because they let it happen on their watch.

The Last Word on Climate Change – Scientist Predictions

As of this writing, our planet is already at one degree Celsius of warming over pre-industrial levels. Because CO_2 is so slow in being reabsorbed by our oceans and plant life, we know that we are looking at an increase of at least .5 degrees Celsius, for a total increase of 1.5 degrees Celsius over pre-industrial levels. That's if we stopped all carbon emissions, worldwide, right now. The Paris climate conference of 2015 pledged to dramatically reduce carbon emissions, and try to keep the planet at no more than 2 degrees Celsius of warming. Not one of these goals agreed to in the Paris conference has been met.

It's believed now that we may reach 3 degrees of warming as early as 2050 and as late as 2100 even with dramatic reductions in carbon emissions. If our world leaders can reduce emissions drastically, then we might keep warming to less than, or equal to 3 degrees Celsius by 2100. If we continue with population increases, fossil fuels and carbon emission increases, then all bets are off. Climatologists don't like to theorize about increases over 5 degrees Celsius, because the results sound like science fiction. However, it's all too real.

This isn't bullshit. This isn't liberal propaganda. It isn't some crazy guy typing on his keyboard and trolling Republicans on Facebook. This is what the vast majority of climate scientists are predicting and HAVE BEEN predicting for years. It's no longer theoretical, either. We are living these changes, right now, as you are reading this page. This isn't conjecture, it's fact.

Here's a list, by increase in temperature, of what we can expect in the coming years. Degrees are in Celsius.

1 degree Celsius of warming. Achieved in 2019.

Increases in climate refugees: In 2019, the Red Cross reported 258 million international migrants in the previous year – the highest number in recorded human history. Although exact numbers are difficult to calculate because many people migrate for a variety of reasons, climate change has been a factor in at least two third of refugees.

Increases in severity of tropical storms, typhoons, and hurricanes worldwide. Examples include:

On March 14, 2019, Mozambique saw the worst storm in her history – Tropical Cyclone Idai. The cyclone damaged 100,000 homes, destroyed 1 million acres of crops, and demolished $1 billion in infrastructure.

On September 1, 2019, the island nation of the Bahamas experienced the worst hurricane in their history. Hurricane Dorian stalled over Abaco and Grand Bahama Islands for over 36 hours, causing unprecedented, catastrophic damage to the residents living there.

Increases in sea level rise: The clearest examples of climate refugees fleeing sea level rise is in the Pacific. The sea level in the western Pacific is rising at the rate of 12 millimeters per year. Islands in the Federated States of Micronesia have dramatically reduced in size, and their freshwater stores are being contaminated by rising seawater.

Wildfires in California, Australia, Europe, and other areas are increasing in size and severity. This does not include the Amazon - yet. Most of those fires are being set intentionally, with the acknowledgement of Brazil's current government, to clear rainforest for cattle grazing.

Glaciers in Greenland, Alaska, and Iceland are melting and calving (the terms for large pieces of ice separating from the mass) at significantly higher rates than ever previously recorded.

Permafrost in Siberia and Canada is becoming less stable, increasing methane emissions from previously frozen, decaying vegetable matter.

Increases in heat waves across the northern hemisphere are becoming more and more common. The United States, Canada, Russia, the entire continent of Europe, Southern Asia, and the Middle East have all reported record temperatures in the last five years. The 18 hottest years on record worldwide have occurred in the last 20 years.

NASA reports that temperature above the arctic circle are increasing faster than previously predicted, AND faster than the rest of the world. The Barents and Kara seas in the Artic have shown a 2.5 degree Celsius warming in the last decade.

In later summer of 2019, Australia reports that the outlook for the Great Barrier Reef is "very poor" after documenting an 89% loss of new coral due to mass bleaching (the direct result of ocean acidification) in 2016 and 2017.

1.5 degrees Celsius warming. Most climate scientists feel that a 1.5 degree increase is inevitable by 2040.

Fourteen percent of global population is exposed to severe heat at least once every five years. Heat stress becomes more deadly as people living in areas without widespread air conditioning are exposed to temperatures above the level that the human body can deal with. The European heat wave of 2019 caused 30,000 deaths, and some estimates are as high as 70,000. This number will pale in comparison to heat related deaths in Europe, the Americas, and Russia in the next twenty years.

Sea level rise will reach .4 meters, or just under 16 inches. This means that all coastal regions, worldwide, will see a 10% reduction in land mass. Worse, saltwater rise means that groundwater in coastal areas will become saline. This means well water will be unusable for irrigation or consumption.

The number of tropical storms will decrease, but the severity will increase. At the time of this writing, meteorologists and weather experts are already discussing adding a Category 6 to the current hurricane scale.

Destabilization in permafrost in mountainous areas starts to cause rock and mudslides. The Andes and the Alps are specifically at risk.

Snowpack in mountainous regions reduces annually on all continents. This directly impacts fresh water volume in rivers during the spring runoff. It also means hydroelectric dams will see a significant decrease in water flow.

Earth's ecosystems start to shift. At this level of warming, 7% of earths land area will change to a new biome. We begin to see significant losses in vertebrates, plants, and insects that lose their range because of biome shift.

The oceans are particularly affected. Warmer water cannot process gas absorption as well as cooler water. We start to see "dead zones" where oxygen isn't present in seawater. Coral reef loss increases to 70-90 percent. Right now, humans get 20% of their protein from seafood. A 1.5 degree rise will cause 1.5 million tons of decline in marine fisheries.

Freshwater availability decreases, as temperatures continue to dry out soil, inland lakes, and rivers faster than pre-industrial temperatures. Areas where drought is common, such as sub-tropical Africa, become even more susceptible to more difficult growing seasons.

Disease in third world countries increases. As the biome changes, insects that are disease vectors proliferate because of longer summers. Deaths from malaria, already one of the leading causes of disease related deaths worldwide, increase significantly.

On a socioeconomic level, climate refugees increase by a factor of 10. Resources in equatorial countries, or countries bordering ever-increasing deserts become more and more scarce. Civil wars, breakdowns in social services, and infant mortality rises.

2 degrees Celsius of warming. Scientists predict a 2 degree increase as early as 2060 and as late as 2100. Change in fossil

fuel policies and consumption is the primary variable in these calculations.

At this point, 37 percent of the earth will see a severe heat wave at least every five years. Deaths from heat related stress reach into the hundreds of thousands annually, even in first world countries.

Sea level rise goes to .46 meters, conservatively. That's 18 inches. Nearly half of the islands in the South Pacific are under water. Land areas built below or at sea level (New Orleans, Miami, The Netherlands, Venice) are all at severe flooding risk.

A feedback loop involving coastal erosion begins to gain momentum. As sea levels rise, more and more coastline is eroded and pulls back into the oceans at low tide, leading to even more loss of coastal land.

Areas of the planet that were susceptible to drought during pre-industrial levels will become impossible to live in. Africa and Australia see significant population losses from starvation.

Killing heatwaves in Europe are an annual occurrence. The countries surrounding the Mediterranean Sea begin to see losses in population from heatwaves as well as starvation from loss of agricultural land.

All the glaciers in Alaska, Greenland and Iceland see dramatic reductions. Sea ice in the arctic will drop to zero at least once every ten years during summer months. Four million square miles of permafrost will thaw (an area the size of the United States).

We begin to see up to 10 percent declines in maize, corn, and wheat harvests on all continents. While some of these grains are not a staple of human diets, they are a staple of poultry and cattle diets. As grasslands begin to die off from lack of water and higher temperatures, the crops are going to be significantly more important to provide feed for cattle and poultry, and thus, protein for human diets.

Living coral reefs drop to 99% of pre-industrial levels due to ocean warming and acidification. As a result, oxygenation of our oceans becomes a major problem. Huge areas of oceans water become "dead zones" devoid of oxygen, where nothing survives. This causes the beginning of another feedback loop – as organic material in the oceans begins to die and rot, more CO_2 is released into the atmosphere.

Third world countries see a complete breakdown of social services. Armed gangs control most water and food throughout the entire African continent. The South American continent, and as far north as Mexico City, sees marked increase in deaths due to disease and malnourishment. The Middle East is consumed in civil wars as heat waves, lack of resources, and starvation become widespread from Egypt to India. The Philippines see widespread overcrowding and overconsumption of resources as she loses 15% of her landmass.

3 degrees Celsius of warming. Scientists tell us that this is possible as early as 2100 and as late as 2150. The variable, as before, is the immediate reduction of fossil fuel use.

At 3 degrees of warming, positive feedback loops begin to accelerate at an exponential pace. If this threshold is passed, then scientists have predicted a "runaway carbon effect", also called a "carbon cycle feedback loop". Simply put: it means that if we reach this level, then we, as a species, lose control of the amount of carbon in the atmosphere.

Feedback loops at 3 degrees of warming:

Permafrost melting becomes an exponential feedback loop. As more permafrost melts, more methane and CO_2 is released into the atmosphere, creating more warming, which creates more permafrost melt.

Bacteria in soils begin to create more CO_2 than plants can absorb. Dying vegetation and soil-based bacteria release more CO_2 into the atmosphere due to warming. This, in turn, increases warming.

The oceans become a runaway feedback loop. Warmer water means less CO_2 absorption. As marine plants and animals die, they create more CO_2 from their decomposition. More CO_2 means warming waters, which means more plants and animals die.

The agricultural centers of the world are completely changed in at the 3 degree threshold. South America and the Amazon basin have turned into savannahs, with annual droughts. Nebraska, Kansas, and Oklahoma return to the dust bowl consequences of the 1930s. Southern India has joined the Middle East in a continent-wide sea of refugees.

Where there isn't drought, there are torrential rainstorms and flooding. The pacific northwest and the east coast of the United States will see longer and larger floods due to extreme rainfall, leading to soil erosion. Spain, Italy, France, and Greece will suffer as well.

Sea level rise floods most coastal cities. Alexandria, Shanghai, Rio, Houston, New York, and London all see significant population displacements. Overall, worldwide, an estimated 300 million people will be forced to move inland.

Most of Africa is incapable of supporting human or animal life. The entire Caribbean is rocked by superstorms that make it impossible to live anywhere between St Kitts and the

Gulf of Mexico. Any cities south of the Arctic Circle and north of New Zealand are broiling morgues in the summer months.

Aid agencies, already at the breaking point, collapse. Climate refugees reach one billion people annually. Governments in the northern hemispheres close their borders to climate refugees coming from the South.

4 degrees Celsius of warming. Scientists estimate as a strong possibility without immediate action. Possible by as early as 2120.

At 4 degrees of warming, Iceland and Greenland are completely ice-free. The melting of these glaciers back into sea ice has cause the rise of our oceans to cover the worlds coastal areas.

Runaway positive feedback loops are in full effect. Most ocean life has either migrated to the poles or is dead or dying. Huge areas of the Pacific and Atlantic are completely anoxic. What's left is covered in floating plastic garbage patches larger than a continent.

There is a complete socio-economic breakdown, worldwide. With no aid agencies left, and northern countries closing their borders, billions of people will be left without food, clothing, clean water, or basic sanitation. There won't be such a thing as refugee camps, as refugee camps typically mean there is a central area for aid. People trying to migrate north will either reach a border and be turned away or shot. Most will simply walk until they die.

Russia and Canada will be the new world powers, with countries like Poland, Sweden, and Ireland having to enforce their borders with deadly force. Greenland will become a major agricultural center.

There will be isolated areas of survivalism throughout Europe and the central United States by people who have prepared for the crisis. The wealthy will survive the mass dying by hiding in their bunkers.

5 degrees Celsius of warming. Scientists rarely discuss this level of warming, but agree this threshold will be breached if we reach 3 degrees of warming, and begin to see runaway positive feedback loops.

As of 2019, our species has increased atmospheric CO_2 to 410 ppm. Pre-industrial levels were around 300 ppm. The last time the planet had this much atmospheric CO_2 was about 3 million years ago. The world's oceans in this time period were between 15-25 meters (50-80 feet) higher than today.

If we get over three degrees of warming, then 5-6 degrees is almost guaranteed because of runaway positive feedback loops. At 5 degrees of warming, atmospheric CO_2 reaches a point of exponential growth. Over time, CO_2 will accumulate in the atmosphere to a point where we are looking at 2000 to 3000 ppm.

The last time the planet was this warm and had this much atmospheric CO_2 was at the end of the Permian period, about 250 million years ago. An accumulation of CO_2, probably due to volcanic activity over 100,000 years, pushed the planet into a runway greenhouse effect. Negative loops occurred – that much volcanic activity would had thrown a huge amount of particulate ash and dust that would have cooled the planet for a time – but once that settled, the greenhouse effect would have been in full force.

Scientist call the end of the Permian period "The Great Dying". The asteroid that killed the dinosaurs 60 million years ago pales in comparison to the loss of biodiversity at the end of the

Permian. Scientists believe that 96% of all vertebrate and marine species were wiped out in this period of time.

If we, as a species, do not take immediate action, then this is our future. It took 100,000 years for CO2 to accumulate to runaway greenhouse levels at the end of the Permian, and we'll probably do it within 250 years.

At 5 to 6 degrees of warming, the majority of the land mass of the planet is simply too hot to support plant or animal life. Only the poles and what is now arctic tundra will be able to support any real levels of population, as these areas of the planet become the new temperate zones.

The majority of the world's oceans are dead. Only the seas around the poles are cool enough to maintain oxygenation. Any coastal regions between New Zealand and the arctic circle are unhabitable from sea level rise, super storms, and worse – poisonous gasses being released from the anaerobic bacteria in non-oxygenated seawater.

As oceans plants and animals die, they sink to the bottom of the ocean and begin to decay. Any available oxygen is used quickly in decomposition, and anaerobic bacteria take over. These bacteria release hydrogen sulfide gas. This is a natural occurrence, but as early as 2004, the planet began seeing coastal areas around heavily polluted countries such as Namibia and other countries seeing mass fish die off from hydrogen sulfide gas. As the oceans die in a 5-degree warming world, any coastal area or island will be subject to deadly clouds of hydrogen sulfide blowing up from sea eruptions. This low-lying gas is lethal in concentrations as little as 50 ppm. Huge clouds of this gas will be brought in on the winds, coming dozens of miles inland and killing all vertebrate life in it's path.

I want to say again – this isn't bullshit. It's not a lone crazy guy making up apocalypse stories on his computer. It's not printed in bleary tracts to pass out at Earth Day festivals. This is fact. This is what will happen if we, as a species, don't take steps to stop it now.

The problems facing us aren't based on how big your SUV is, or the importation of Brazilian beef. It's not about leaving your lights on or off overnight. It's not about a carbon tax on large companies. These problems are all smaller issues around one big issue, and that is: **There are too many people on the planet, using too many resources.**

We can address the problem in one of two ways. The first is to allow nature to take her course. Lack of resources and a changing biosphere will cause the human die off I've predicted in these pages. We will see a significant loss of human life, and by significant, I mean 25% on the low end, and 75% on the high end. A loss of 75% of the earth population means 5.6 billion people would die.

The other option is that we can prepare, plan, and mitigate the die off. We will still see a global loss of life – far more than anything that has ever happened to the human race in our recorded history. However, with some forethought from our leaders, we may be able to keep warming to a minimum, and the human losses to only a few hundred million.

Chapter 5: The Last Hope - How to Slow Climate Change by Changing Population Levels

In the majority of books, articles, and websites I've read in researching this book, people talk about ways to reduce our "carbon consumption" or our "carbon footprint". That's nice, but given the choice between driving a comfortable car, having hot coffee, and hot water for bathing, people are always going to choose what is most comfortable for them. I'm no exception – as I am writing this right now, I've got ice in my drink, a refrigerator to store food, a car in my driveway, and lights on all over the house.

In addition, I've read several articles on using some sort of atmospheric gas, intentionally released into the atmosphere, to reflect sunlight and therefore, lower the temperature globally of the planet. I'd like to go on record here and say that this is the worst idea of all possible bad ideas. Changing the atmosphere of the planet is what got us here in the first place. Intentionally changing the atmosphere on a global level will have a variety of unintended consequences, and they are all likely to be really, really bad.

Consider the cane toad. In 1935, a group of well-meaning scientists released about 3000 cane toads in Queensland, Australia. This non-indigenous amphibian was brought into the sugarcane plantations of the area to try and control the destructive cane beetle population. The cane toads weren't all that great at controlling the beetle population, but they were remarkably successful at breeding, destroying the natural fauna of the area, poisoning native species, and killing the natural predators of a variety of insects, causing the insects, also, to breed almost completely unchecked. The Australian government now estimates somewhere around 200 million cane toads on the continent, and the cane toad is continually referenced as one of the worst biological events to ever happen to an indigenous habitat. This is just one example of how well-meaning people have really screwed up things for the rest of us. Once more, with feeling – changing the atmosphere to fix the warming problem is a bad idea.

The obvious answer I hardly see anyone talking about is to reduce population levels. This allows events such as reforestation to occur; it lessens the feedback loops that are a part of climate change; and it changes the balance of the consumption of natural resources.

The biggest argument to population controls is the argument for the "sanctity of human life". People don't seem to realize that this is a very, very new thought process. Nobody really cared about the sanctity of human life up until about 150 years ago. The Vatican certainly didn't care – for the last 1500 years or so, their policies were aimed at increasing population levels of devout Catholics, while expunging heresy at the point of a sword. The Roman Empire didn't recognize a period of mourning for a child less than a year old. In ancient Greece, infants with deformities were destroyed immediately after birth. I certainly don't have to bring up the caste system that has plagued Asia for the last thousand years.

Am I saying that human life isn't important? Of course not. I'm also not arguing for eugenics, selective breeding, ageism, or racism. While I do believe that genetic manipulation will produce "designer babies", that's a topic we'll cover in the next book. For now, all I'm asking is for a little bit of latitude. For just a moment, I want you, the reader, to consider the greater implications of a mass die off versus the possibility of voluntary population controls.

We already know that the ruling class has the foreknowledge of a mass die off. At best, they are preparing themselves for the mass die off by tunneling underground bunkers and storing supplies. At worst, they are *planning* on a mass die off *while* tunneling underground bunkers and storing supplies. We already know that pure logic suggests a mass die off is the only way to save the planet from global warming. What if we are faced with the specter of a mass die off of epic proportions versus the voluntary controlling of population though policies designed to achieve a lowering of carbon in the atmosphere. Is this something we can open a discussion on?

Maybe so, maybe not. I really fear that modern governments are like the Titanic – too big to turn in time to avoid the inevitable collision. Additionally, I will wholeheartedly agree that without the support of China and India any discussion on climate change is doomed to fail. China is, without a doubt, the largest polluter in the history of the world in terms of both toxic pollutants and CO2 emissions. India is the third largest. Can you guess who is the second largest polluter? (Hint: It's the United States.)

Furthermore, it seems that any idea that might have a negative influence on ANY group of people is met with shocked indignation, or worse, a label of "racist", "bigoted", "misogynistic", or "misandristic". It's so much easier to place a label on an idea than it is to actually think about it. After all, if you can put a label on something, no matter if that label is true or not, it's easier to compartmentalize and dismiss – especially in our world of 10 second news bites.

On the other hand, I firmly believe that the definition of 'whining' is to complain about a problem without offering a valid solution. So, in that train of thought, here's my list of ideas of how to reduce our carbon footprint. Some of these will be unpopular. Some of these are based on historical events, some I've gathered from other sources, and some are my own ideas. I'm sure the collection of them will cause a lot of shocked indignation, and even some labels. If so, try to ignore the label and look at the idea. Maybe it will lessen the impact of the coming die off. Who knows? Maybe some of these ideas, if put in motion, will have a positive effect on yourself and your family.

Population Changes from Conception to End of Life

For a moment, let's talk about population control from both sides of the arc of the human lifespan – from the beginning, and from the end.

Abortion

Abortion. Why are we still even talking about this? I'm a white, middle aged male and I don't understand the amount of column inches that are still being used to discuss it. There are only two reasons I can see to continue the argument – to punish women for having sex; or, to deflect the news away from other, more important events that are happening in the world today.

Women are responsible for their bodies in all other aspects of the law. If a woman wants to get an abortion, she should be able to get one safely, with a minimum amount of bullshit. It makes no sense to force someone to have a child. The child become a burden on an already overwhelmed system. Out of all the abortions performed in the world annually, 92% were considered an unintended pregnancy. I'm always amazed at pro-life people. Are you going to raise the child of an unwanted union? No? Then hush. Allowing women full access to safe abortions is not only the right thing to do, it also reduces nearly

42 million unwanted people in the world annually, including the 650,000 in the United States.

In fact, I'll go one step farther. I believe that women's health, and specifically, contraception, should be available free of charge through government funding. I think that a woman should be able to go into a state clinic, or a women's health clinic funded by the state at any point in time after she turns 16. I don't agree that government should pay for abortions – I think that leads to abortion being used as a form of birth control. But I do believe that both women and men should be able to receive birth control, condoms, and other forms of contraception at no charge. I think these same clinics should offer free STD testing and treatment for anyone, regardless of their health insurance. For the purposes of our arguments against overpopulation, I think it's an incredibly inexpensive way of maintaining a healthy population.

Birth Planning Programs

In 1975, China adopted the slogan "Later, Longer and Fewer" and urged its citizens in urban areas to have no more than two children, and rural couples to have no more than three children. Then, in 1979, the Chinese Government limited couples of the Han ethic group – the ethnic majority in China – to one child per household.

Why did China do this? China saw its population exceed 800 million in 1970, and the State Council mandated sharp reductions in population growth rates. (By comparison, the US only has 327 million people living within its borders today, and only had 205 million people living in its borders in 1970.) China was rightfully concerned about natural resources and the availability to feed its citizens. After the Great Leap Forward and the resulting famine, China entered its own baby boom, similar to the one the US experienced after WW2. China felt that limiting population growth was the key to making sure everyone had access to food, housing and clean water.

The one child rule was unpopular in the rest of the world because of human rights questions. In China, as in most other societies, a male is favored over a female newborn. The incidents of selective abortion skyrocketed – a couple would perform an ultrasound to determine sex, and terminate the child if it was female. China passed laws outlawing sex selective abortions in 2005, but the law was difficult to enforce. This has led to a significant difference in sex ratio in China than that of the rest of the world. In China today, there are an estimated 36 million more men than women.

To enforce the policy, China offered incentives to families who followed the rule, and punished couples who flaunted it. "One child certificates" were issued to couples that followed the rule, allowing them better child care, housing, and longer maternity leave. China also hired more than a million part- and full-time workers to encourage women to use birth control, and to harass the ones that didn't.

Did it work? If you look at the stated objective of the policy, then yes, it worked well. China has been able to focus their citizens and resources into one of the most prolific manufacturing countries in the world today. It's been estimated that between 400 and 500 million births were avoided between 1975 and 2019. In 1984, China began easing some of the restrictions, and did so again in 2006 and 2013. In 2015, China publicly announced that all couples would be allowed to have two children. The government has said that the one-child policy allowed countless families to rise above poverty by easing the strain on the country's limited resources, and it is inarguable that China now provides more manufactured goods than any country in the history of the world.

But in the 35 years of one-child restrictions, a change in the idea of what a perfect family changed as well. Now, it's not just government mandate – two generations of one and two child households has led young adults prefer smaller families to larger ones. This is going to be a problem in China as the population born after the Great Leap Forward, but before the one child mandate was enacted, gets older.

An aging population is the largest concern with this form of population control. China is now concerned that there won't be enough young people to care for it's elderly – which is a problem the United States and most of the rest of the developed world will face as well. China even has a name for it. It's called the 4-2-1 problem, when one child is forced to care for an aging set of parents and two elderly sets of grandparents.

What lessons can we learn from China? Could other developed nations enact similar legislation and possibly curb the effects of climate change by limiting population growth? I think that it's possible, given the alterative.

In fact, I think it dovetails quite nicely with my previous section on abortion, women's rights, and the government funding of sexual health for both men and women. The national average of children per married couple in the United States is about 1.9 children per household. The last time it was over 2 children per household was in 1977.

I'm not saying that we should allow government funded abortions. I'm saying we allow birth control and sexual health to be a human right. If a couple wanted to have more than two children, I think that's wonderful. Unlike communist environment of the Chinese, I don't think it's constitutionally possible to limit people to a certain number of children in the United States.

However, I do feel that more than two children per couple is a burden on our system of education, our natural resources, and healthcare system. Over time, it will also be an added burden on the sexual health clinics I've talked about here already. I think that an added tax on couples with more than two children would be in order. If you are willing and financially able to have more than two children, that's great. If you'd rather wait and have your third, fourth, or fifth child after you have the ability to pay the tax burden of additional children, that's great too. I think this makes a lot more sense than a carbon tax, and it's a lot harder to loophole as well, for obvious reasons.

There is a glaring problem with this idea. Right now, in the United States, poor people have more children per household than rich people do. In 2017, there were 66.44 births per 1000 women under an income level of $10,000 per year. Wealthy women that made over $200,000 per year had 43.92 births per 1000 women. Obviously, you can't tax people at less than poverty income levels at a higher rate when they can't even pay for the taxes they have now – much less for the children themselves.

My answer to this concern is this: poor people don't want more children than rich people – they just don't have the ability to afford birth control, doctor's visits, family planning, or abortions. If you level the playing field between high income and low income families by providing sexual healthcare at no charge, then you automatically even out the amount of births per capita, regardless of household income.

There will still be pundits who feel that a higher tax on families with three or more children is unfair, or even "Un-American". But is it any more or less fair than a carbon tax? An inheritance tax? A luxury tax? It is any more a tax on the poor than a state lottery, or scratch off tickets? I don't think so. I think a straightforward conversation about state-sponsored sexual health and population controls is in the best interests of the long-term health of our Nation.

End of Life

Imagine, for a moment, if modern medicine was able to cure cancer, heart disease, emphysema, and strokes with a single pill. Found a lump in your breast? One pill, and you're cured. Shortness of breath? Sudden pain in your chest? One injection, and it's over – no pacemakers, no transplants. Better yet, one preventative pill could cure all four, for the rest of your life. Wouldn't that be wonderful?

Actually, it would be a catastrophe of biblical proportions. Right now, the life expectancy in the United States (and the rest of the developed world) hovers at around 79 years. If we could cure the four main causes of death, it would jump life expectancy by 20 years.

In the US, annual death from these four causes reaches about 1.5 million people per year. Worldwide, it's 30.2 million. The worldwide population increase by birth is 82 million people per year. We would increase world population by a third annually – for the next twenty years.

The United States healthcare system, already arguably on the brink of collapse, would implode in five years. Not only would there be too many people for our existing healthcare providers to treat, the vast majority of the people we've saved would be elderly, which would put an even heavier burden on our healthcare system. Medicare would go bankrupt. Social Security would go bankrupt. The coming die off would be moved up by 15 years, and the total number of dead would be increase by a third.

Of course, the governments of the world know this. Of the many conspiracy theories I've read while researching this book, one of the few that makes actual sense is that a cure for cancer has been developed, and that world governments are suppressing it. Cancer alone accounts to 9.6 million deaths per year, annually. Adding this amount of people back onto global population would be a significant drain on existing resources.

Event without a magic pill that cures the top five causes of death worldwide, we are still looking down the barrel of a significant aging problem. By 2035, the US Census Bureau anticipates that people over the age of 65 will outnumber people under the age of 18. This is the first time in history that this has happened, and as we've discussed, it's going to be a problem across the developed world. As more people are aging past 65 years old, the strain on healthcare, mental health treatments for the elderly, and basic necessities such as housing, food, and clean water are going to put a strain on existing infrastructure.

Another major issue with an aging population is wealth retention. Most people don't realize it, but even on a global scale, wealth is finite. There's only so much wealth, worldwide, to go around. An aging population stores wealth, a younger population spends it. By storing wealth, economies stagnate. If you reach a point where more of a nation's population is over 65 years old than there are people under that age limit, then you reach a point where more wealth is being stored rather than being spent in an active economy.

Here's my idea of a solution, for better or worse. I think that all people over the age of 55 who are accepting Medicare or Social Security should be ineligible for resuscitation after a life threating injury or medical event. I'm talking about a government mandated DNR order for anyone over retirement age. I would also consider legislation that would block any recipient of Medicare or Social Security over the age of 55 from accepting organ donations, whether donated or lab-printed.

That's probably the hardest paragraph I've ever written. My parents are still alive, and I love them deeply. And talk about a political bomb! Who is going to write that legislation?

However, consider the implications. You'd see an immediate increase in funds available in Medicare. An immediate increase in tissues available for implantation. You'd also see an immediate change in population numbers, which would increase the resources we have available to the youth and to the working populations, and immediately lower the amount of CO_2 being pumped into our atmosphere.

I think that everyone will agree that, given the choice, a healthy kidney for implantation should go to a young person with genetic renal disease, vs an older person with a history of smoking, obesity and diabetes. A more difficult question, but one that I think would have 90% of the population in agreement would be a hypothetical situation where a child that is involved in an auto wreck should be given first priority for a place in an ICU room versus a healthy senior, critically injured in the same accident.

However, agreement for other points is going to be harder to achieve. Imagine a woman in her late 60's who has a heart attack. With current life expectance rates, if resuscitated, she could expect another 10 to 15 years of life. She's a grandmother, has led a productive, healthy life, with children, grandchildren, in-laws – an entire extended family. On the other hand, she is a drain on existing resources. As a non-working senior, the medical care, housing, food, and water she will consume could be used for other, younger people. More importantly, her carbon footprint is a contributor, although an infinitesimal one, to the climate changes that are precipitating the coming die off. Now, take away the hypothetic aspect of the question and imagine she is your grandmother. How does that change your perception of the situation?

Am I saying that we need to stop preventative medicine? Absolutely not. And there are several obvious loopholes here – people looking for tissue implantation could easily go to other countries to achieve these types of surgeries. Anyone with the means to retire without Social Security or Medicare could easily simply pay for resuscitation measures.

However, immediate steps must be taken. It's not enough to change our habits. We have to take a serious look at population numbers as well. Even with these changes, I feel a human die off is inevitable. Some would argue that changing abortion policies, or end of life policies is worse than an increased number of people in the next human die off. After all, we get to keep our morality – or maybe our sentimentality – by not making these changes. But is it worth the pile of corpses when the inevitable die off comes?

Policy Changes that Reduce Population and Increase Conservation

We've talked about some population control options that will help us reduce the carbon footprint of our population, and allow significant increases in resources. Let's talk now about some policy changes that will reduce population numbers, and will hopefully increase resources and infrastructure as well.

Death Row Inmates

We've talked about the US prison system before, and the incredible levels of incarcerated people there are in the United States today. I think it's an unbearable weight on the backs of the poor and lower middle classes that non-violent offenders are incarcerated by for-profit prisons. However, I'd like to take a moment and talk about people that are on death row – people convicted by a court of law and sentenced to death.

The United States is one of the few countries left in the developed world that still practices the death penalty. We are the only country in North America that does it – both Canada and Mexico abolished it. Most of the EU has abolished the death penalty as well.

I think that, given the coming human die off, and the diminishing amount of resources we have available to us today, that anyone sentenced to die should have their sentence carried out. Any new person sentenced to die should have their sentence carried out within 30 days of receiving the sentence.

Yes, I understand that an estimated 4% of the inmates on death row may have been convicted erroneously. Yes, I'm sure that I would feel different if I was the guy sitting on death row. But let's face it – of the 3125 people on death row in the United States, most of them will wait 15 years between sentencing and execution. In the continuing theme of conserving resources by reducing population – this is the easiest choice.

Physician Assisted Death

To me, this one is right up there with abortion rights. If we are talking about population reduction, then why wouldn't you allow a mentally stable person, who doesn't want to continue living with a debilitating or painful medical condition, the right to check out with some dignity?

Oregon leads the US with physician assisted death, with 168 deaths in 2018. Other states have physician assisted dying statues on the books, including California, Colorado, DC, Hawaii, Vermont, Washington, and Montana.

Physician assisted death is legal in other countries in the developed world: Canada, Belgium, Colombia, The Netherland, Luxembourg, and of course, Switzerland allows for physician assisted death. Switzerland has become known as a destination for suicide tourism – that is, people who want to end their lives but are not able to do so in their home country. As the world's population ages, this is going to be more and more of a solution for people who are going to want to end their lives by choice.

Addiction and Drug Use

Drug abuse is rampant in the United States for a variety of reasons. Once of the most prominent reasons is the profit margins to Big Pharma, which drove the intentional and willful distribution of the incredibly addictive painkiller Oxycontin, which, in turn has driven the opioid epidemic we are currently in the grip of. But to be fair, it's not the only reason that America has more overdoses than any other country in the developed world.

In 2017, there were 37,133 deaths related to car accidents. There were 70,200 overdoses. That means you are nearly twice as likely to die from overdose in the United States than you are from a vehicle accident.

It's true that our nation's leaders didn't respond to the opioid epidemic as quickly as they should have. (Maybe they should have read their history.) It's also true that a big reason they didn't is because of the lobbyists and donations made in favor of Big Pharma to our nation's lawmakers. But the abortive War on Drugs is also a major factor, as well as our prison system. By criminalizing an addict, you've made them an outcast.

Some of our nation's communities have recognized this and have begun making legislation that protects users. Needle exchange programs are available. Good Samaritan laws are another promising step towards protecting someone who calls in an overdose. States with Good Samaritan laws offer limited immunity to someone that calls in an overdose. However, there are states that do not have these laws, and as a result, other users that might be with the person overdosing are reluctant to call medical personnel or police. For example, in Texas, if you call in an overdose, then you are automatically suspected of being a user yourself, and are taken to jail.

Most people don't realize that this epidemic is actually the second time that the US has been gripped by an addition to opioids. The first opioid epidemic in the United States lasted for nearly 50 years, nearly the exact timeframe we've had the current War on Drugs. Once again, we'll look at some history to talk about our path for the future.

The first opioid epidemic began in the United States during the Civil War. The Union Army alone issued nearly 10 million opium pills to it's soldiers, plus an additional 2.8 million ounces of opium powers and tinctures. When the Civil War ended in 1865, veterans returning home from both sides of the war had injuries that would last them a lifetime, and an unknown percentage were already addicts when the war ended. (In fact, morphine dependence after the war was nicknamed "Soldier's Disease".) With the advent of the hypodermic syringe in 1856, doctors used morphine intravenously to treat everything from major trauma to menstrual cramps and even morning sickness in pregnant women. Morphine was used routinely to quiet teething children. Laudanum was used in everything from sleeping tonics to headache remedies. Since it was sold over the counter, anyone could buy it. By 1890, it's estimated that 1 in 200 Americans were addicted to morphine.

But public opinion began to turn. Although it's impossible to get reliable cause of death information from this period of American history, statistically, overdose deaths would have affected hundreds of thousands of people – similar to today. People began questioning their caregivers and lawmakers, and the drug companies that were promoting and profiting from drug sales. (One of the largest distributors was a small company based in Germany called Bayer). State laws passed between 1895 and 1914 restricted sale of opiates to people with valid prescriptions. In 1906, in response to a Collier's magazine article on the deceitful practices of the patent medicine industry – that is, drug manufacturers – Congress passed the Food and Drug Act. For the first time in American history, medicines had to be sold according to purity and strength, and ingredients like opium, cocaine, and morphine had to be listed on the label. In 1914, the Harrison Act, signed by Wilson, required anyone selling opiates or cocaine to register with the federal government. Finally, in 1924, the United States made heroin illegal, effectively ending the opioid epidemic of the time.

What did the United States do for people who were addicts after the 1924 law that made heroin illegal? The answer is – very little. Several cities opened treatment centers in the years after the Harrison act was passed, but Wilson's Treasury Department stomped them out of existence by 1921. Most of these clinics were treatment centers that allowed older men and women who were long time addicts to maintain their habit. But younger addicts were pretty much left on their own to either stop using, or... not. The fact of the matter is that addicts in this time period are very similar to current addicts. They did one of two things – overdosed, or got clean.

Interestingly, there are several parallels between the opioid addition of the late 19th century and the one we are facing today. Women are more affected than men. White people are more likely to become addicts that other minorities, particularly blacks – which, ironically, is most likely due to subtle racism against black people by medical personnel. Also, the timeframes are almost identical. Nixon first declared the War on Drugs in 1971, and public opinion against Purdue (the makers of Oxycontin) turned in 2007, when the company pleaded guilty in federal court to fraudulently marketing their drug. That's 36 years. If we look at the Collier's articles in 1906 versus the end of the Civil War, that's 41 years.

You also must consider the curve of public opinion. In 1996, as in 1865, the 'miracle drug' would have had a high public opinion. In both cases, initial use saved people from pain. But as the negative side of the drug comes out, public opinion wanes. Watching a family member descend into addiction is heartbreaking. Watching a child die from an overdose is shattering to parents and family members. Finally, when you are watching your third (or fourth, or fifth) friend twitching on the premanufactured floor of your singlewide, public opinion really takes a nosedive. We are now at this stage of our current opioid epidemic. It took 20 years from this point in the epidemic of the 19th century to come to a close. Hopefully, it won't take that long with the current epidemic.

Out of all the topics of this book that illustrate problems facing our nation, addiction is the hardest to address. We have spent the last fifty years criminalizing drug users – about the same amount of time the first opioid epidemic gripped the United States. Now, as before, the opioid epidemic has shown that anyone can become an addict. Construction workers, high school football players, our veterans, people with chronic pain – children, mothers, fathers, brothers, sisters – some died because of their own choices, but the majority of overdose deaths in recent years were from normal people who became addicts not by choice, but because they were misled.

The biggest difference between the opioid epidemic of the late 1900's and the one we are facing today is greed. In the 19th century, morphine, laudanum, and later, heroin, were treated as miracle cures – medical science didn't know that they were creating addicts until the late 1880's. In this modern time, Big Pharma knew they were pushing a substance that was addictive after less than 10 doses. Our lawmakers knew. When they write today's history, they will write that people were misled by their caregivers and lawmakers, and became addicted to a drug that everyone told them was safe.

On the other hand, pure logic tells us the next die off will include normal people as well. Addicts are a drain on the system. They use more resources that people that are not addicts. An addict is more likely to steal, damage property, and put their lives and the lives of others at risk. We've already learned how the United States dealt with addicts after the first opioid epidemic. It was a harsh choice, but ultimately, was good for the country.

I've known people who were drug users and were able to stop. I've known people that were users that stopped for a period of time, but were drawn back into that world from their own personal weakness. No matter how the person became an addict, they have to be able to be strong enough to break their own cycle and get clean again – and stay that way.

I believe we should help the people trying to get clean. I believe it should be a part of our public health system to help people get clean. But for the purposes of this book, a serial addict is a drain on the system. There has to come a point where the addict makes their own decision to either get clean, or remove themselves from the equation.

If we are looking at reduction of population numbers, then these three policy suggestions have relatively paltry return on investment. The use of the death penalty, physician assisted death, and the tragedy of addiction are well below the millions of lives that could be affected by the population controls we discussed in the beginning of this chapter. Even if these ideas were made into policy, we are still looking at the imminent death of millions of people due to climate change. On the other hand, maybe there is a better solution.

As I've said before, I firmly believe that someone who complains about a problem without offering a solution is whining. With that thought, I'd like to offer a little hope – a proverbial ray of sunshine. For just a moment, I'd like to talk about some policy changes that might help us avoid the die off predicted in the rest of book. Some of these are my personal ideas, and some I have reviewed and revised from other authors and futurists like myself. Other people – better people – have offered ideas that would keep us from the die off predicted in this book and have been ignored. But maybe, just maybe, someone will read these next pages and act on the ideas written here. History will tell.

The Green New Deal vs The New Deal of 1933

In February of 2019, Representative Alexandria Ocasio-Cortez and Senator Ed Markey (both Democrats) submitted legislation called the "Green New Deal". The document was a non-binding resolution, designed not to create any new programs, but to put on the Congressional floor that climate change and related issues needed to be addressed in the coming years. It's a fascinating document that was the first step toward having a domestic, legislative conversation about what we can do to address climate change and the carbon footprint of the individual. The Senate failed to pass a procedural vote on the measure, something that didn't really surprise anyone. It wasn't supposed to pass. It was designed to open a discussion, and it certainly did that – the Republican party did everything but burn copies of the bill on live TV.

I've read the Green New Deal, and I think you should too. I've included it in the first appendix of this book so you will. The only real problem is, it won't work.

There are a few reasons it won't work, and most of them have roots in the original New Deal itself – the series of measures instituted by Franklin Roosevelt to get America out of the Great Depression. As I'm sure you can guess by now, we're going to look at a little history to gain some insight into the future.

Franklin Delano Roosevelt, a Democrat, was elected President in 1932, just three years after the stock market crash of 1929 that began the Great Depression. He beat the previous President, Herbert Hoover, in a landslide. Many Americans at the time viewed Hoover to be at fault for the Great Depression, and although he wasn't, the conservative financial policies he enacted during his Presidency didn't (and couldn't) bring any relief from the poverty, joblessness, and debt the average American was being crushed under. Additionally, the common man was sick and tired of the Volstead Act, and one of the main platforms of the Democratic Party was the repeal of Prohibition.

The Republicans had little chance of winning the 1932 election. It was exceptionally bad for Hoover, whose campaign appearances were disastrous – he often had objects thrown at him or at his vehicle as he rode through city streets. Imagine that in today's world!

FDR was arguably the greatest American President of the 20th century. He took the reins of the country in the depths of the Great Depression, built innumerable public works projects, began the FDIC, set the first minimum wage, and became a wartime President when the United States entered WWII after the attack on Pearl Harbor. FDR served four terms as President – the most ever served in the history of the US.

But most people today don't realize that FDR was not universally loved. He did more to increase the middle class than any other President before or since, but Republicans, specifically the wealthy conservatives that enjoyed the status quo, hated FDR. His policies of fair employment benefitted several major minority groups, which was a major sticking point for the wealthy white businessmen. FDR also signed legislation that established a minimum wage, introduced the 40-hour work week, eliminated child labor, and guaranteed overtime of 1.5 times the hourly rate for most jobs. These policy changes cost the wealthy billions of dollars.

Why the "Green New Deal" and Similar Legislation Simply Won't Work

In 1933, as we've discussed, nearly 50% of the population of the nation was rural. Government was "for the people, by the people". That's simply not the case in America in our current time. As we've stated before, government now is "for the corporation, by the corporation". The first reason that the Green New Deal will never pass, and that similar legislation has little chance of passing, is that the Congress is now for sale, and the corporations with the most money like the status quo.

The second reason is Roosevelt himself. FDR was a born negotiator, a masterful politician, and served as a public servant in two of the nation's greatest challenges of the 20th century. In his first Presidential campaign, FDR campaigned on two major points – the relief, recovery, and reform of the Great Depression, and the repeal of Prohibition. In his first 100 days in office, Roosevelt enacted and signed into law the major portion of his New Deal legislation, as promised, and Prohibition was repealed in the same year he was elected.

Can you imagine a Presidential candidate in modern times who campaigned on a series of promises, and had the majority of them signed into law within 100 days of his Presidency? A candidate that was able to push through a constitutional amendment within the first year of his Presidency? Understand that history has proved Roosevelt right, but in 1933, his policies were completely untested and seen as reckless and dangerous by many people. Granted, Hoover and the failure of the Republicans to ease the strain of the Great Depression helped him, but it was Roosevelt himself that convinced Congress, and the Nation, that he was right.

I don't believe either the Republicans or the Democrats have a candidate that could accomplish anywhere near as much as

Roosevelt did in his first 100 days in office. This is the core of the second reason climate change legislation is hobbled in our nation right now – we don't have a leader who could get both sides to agree on the major changes and sweeping legislation that it would take to bring climate change on the floor of the Congress. Or maybe, we just can't find a leader who has the courage to stand up for what is right, not just for the company that puts the most money into their Super PAC.

Finally, the third reason climate change legislation won't be passed is because it has no direct effect on us as a population – yet. In 1933, most of American's were struggling under the Great Depression. Nearly 25% of American's couldn't find a job, and those that could saw their wages frozen or lowered, or saw their hours reduced to part time. Even high-profile professions such as doctors and lawyers saw their wages reduced by 40%. Nearly all Americans made do without luxuries that were common in the decade prior.

How has climate change affected the life of the modern family? There have been isolated storms where people have lost their homes and their lives, but climate change disinformation sources are quick to label them as 'normal' storms, not severe weather worsened by climate change. Worse yet, these same disinformation campaigns tell middle class people that alternative energy sources costs Americans their jobs. These pundits continually state that climate change is false, and that reducing the carbon footprint of the middle class will actually cost more money and time, and cause more stress in their lives.

In short, there's no real reason that the average American would embrace climate change legislation. It has no effect on their daily lives. Sure, it's a bummer to think about dying polar bears, and the Great Pacific Garbage Patch, and how plastic bags are killing the whales. But does it affect getting your kids to school on time? Does it affect your food bill? Does it affect your ability to take a vacation one week out of the year?

The answer is – not yet. But it will affect the lives of the average person, and soon. Ask yourself, could Roosevelt have pushed New Deal legislation if the American public hadn't been under the financial stress of the Great Depression? The fact is, even as pivotal as FDR was, he couldn't have enacted the policies of the New Deal without the struggles the public were going through at the time.

Climate change simply isn't causing any real problems for the average person. However, the ignorance of climate change science and the ignoring of climate change legislation by the American government and the other governments of the developed nations will cause the next human die off. Once it starts affecting the average American, then maybe we will be able to enact climate change legislation. By then, however, it will be far too late.

A New Conservation Corps?

Just recently, I took a trip with my kids to West Texas, and we did a canoe trip on the Rio Grande. I noticed the water levels were low and asked our guide about them. "Oh." he said off-handedly. "That's because of the illegal irrigation that the farmers on the Mexico side are doing." All up and down the Rio Grande, people are pumping the water out of the river to a point where it is literally running dry.   I'll grant you that this is a small example of a much larger problem, but truly, aren't they all? Most importantly, this example highlights the multinational issue that is climate change.

As we've discussed before, climate change is a global problem and it's going to require a global solution. To truly address climate change, the nations of the world will need to come together and agree to a series of technological and budget mobilizations. It can't be like the Paris Agreements of 2015, or the Kyoto Agreements of 1997. These agreements worked in theory, but they were unenforceable. There were no consequences for not meeting the deadlines or emissions thresholds. As a result, not a single country that signed the Paris Accords has met the goals of the document – not one.

Maybe we, on our continent, can start small and show the world a way it can be done. Maybe if the residents of the United States work with our neighbors to the North and South – Canada and Mexico – we can build a partnership that begins to reverse the effects and brings a level of clean technology and energy to the people living on the North American continent.

I'd like to see an initiative – let's call it the North American Continent Initiative – that brings the needs and resources of the three countries on the continent together. This coalition would use the material and personnel assets of the three countries together to provide a workforce whose sole purpose would be the building of renewable power infrastructure throughout the North American continent.

The organizational structure would be a governing body representing all three countries. Each country would appoint three positions – a person in charge of the available real and material assets of the country, a person in charge of the personnel assets of the country, and a person in charge of the projects needed to be addressed in each country. These three representatives would be appointed by their respective governments.

The North American Continent Initiative would have to be an organization that worked independently of Mexico, the US, and Canada. In effect, it would become a quasi-military force funded by all three countries, for the benefit of the residents of all three countries.

Imagine for just a moment – an organization whose sole purpose it is to generate clean water and renewable power across the entire North American continent. This organization would not only provide renewable resources for residents of all three countries, but would also provide training for anyone willing and able to work.

- The American military has several decommissioned bases already built with barracks, machine shops, administration building, and airfields. We could easily convert these spaces to be utilized to train and equip a multinational Conservation Corps.

- We could fund the construction of water desalinization plants along the entire Pacific shoreline dedicated to providing clean water for irrigation. Instead of oil pipelines, we build water pipelines to restore our rivers and subterranean aquifers.

- We could train an entire generation of workers on how to build solar panels, install them into solar farms, and provide inexpensive energy to people in populated areas without a single penny being spent by residents.

- We could build charging ports for electric cars, trucks, and semis at the intersection of every major highway in the United States, Mexico, and Canada.

- We could reforest the entire North American and South American continents. Rather than clearing forests for agriculture, we could provide training on vertical farming and methods of livestock management.

- Once the Initiative is up and running, we accept people from other countries on work/study programs. Developing nations would be given priority. Cooperation in nations, especially nations in northern African, Southeast Asia, and the Middle East would be given priority and incentives to participate.

Staffing an endeavor like this will be challenging, so I propose that we ignore citizenship. Any person would have the ability to join the Initiative, as long as they are not felons in their host country, are willing to learn a common language, have an baseline level of intelligence or skillset, and be willing to work a minimum of four years in the Initiative. Employment opportunities will be in everything from installation all the way up to R&D of emerging technologies. A person in the Initiative will not have citizenship in any one country – they will be citizens of the Initiative until discharged. The initiative would have their own passports, security, and court system for it's members.

The obvious advantage here is a multinational, self-governing force specifically designed to combat climate change. This group wouldn't be funded by corporations, so it doesn't have to answer to them or follow their desires to defraud the common man. Although backed by the military, it's purpose is not to use military might to coerce people or topple governments. The purpose is to show the world how we can come together and combat climate change, feed the hungry, and train people who are willing to learn.

I realize that people will read this and think I am naïve. However, I have to believe that feeding children makes you a friend to their parents. Killing children makes you their enemy. Preserving the world for our children makes you a hero. Destroying it makes you a villain. Yes, I do think it's that simple, and I think the resources to staff and fund this type of organization are already there – we just need to change our minds on how it's being used.

Why are Hispanic Immigrants a Problem – and Not a Solution?

The United States has gone through periods of acceptance and denial of different ethnic groups during its 250 year history. While there is no question that we are a nation of immigrants, there have been plenty of times in our history when a certain group of people have not been welcomed into the melting pot that is the United States gene pool.

There are several examples of this type of racism in our nation's recent history, and some that are still used in our vernacular today. The term "shanty Irish" originated as a derogatory term for the tens of thousands of Irish that emigrated to the United States during the Great Potato Famine. You've probably heard someone say that "They got gypped on that deal" or heard someone call a bad business arrangement "a dirty gyp". The origination of that word was "Gypsy", a term used to define the thousands of Romanians that emigrated to the United States in front of Hitler's occupation of Romania in the early 1930's.

However, the longest lasting and most violent example of xenophobia was probably the "Yellow Peril" of the 1870's to the 1920's. Miners during the Gold Rush in the American West hated Chinese immigrants, who worked almost exclusively on the railroads. On September 2, in 1882, a riot broke out against Chinese workers in Rock Springs, Wyoming. In the violent exchange, at least 28 Chinese immigrants were killed in clashes with the white miners, none of whom were ever charged with a crime. In fact, most were treated as heroes for attempting an "ethnic cleansing". The Rock Springs Massacre wasn't the first or the last instance of violent activity towards Chinese immigrants, but it was the most publicized, garnering editorials as far east as the *New York Times*. Most of the printed editorials of the day were actually in favor of the white miners.

The Founding Fathers included in the Constitution the power for Congress to establish a uniform rule of naturalization. In 1790, the first naturalization law was passed, enabling "those who had resided in the country for two years and had kept their current state of residence for a year" to apply for citizenship. Although immigration laws have been addressed, enacted, and revised several times in our nation's history, the highlights are the passing of the 14th Amendment in 1868 to allow citizenship to former slaves, the Chinese Exclusion Act in 1882 that limited further Chinese immigration, and the 1921 Emergency Quota Act, which limited people by race based on the 1910 census. It wasn't until 1965 that the racial based system of 'national origin quotas' was abolished in the United States – which, ironically, was signed into law by Johnson.

What's funny about the modern immigration issue – that is, Hispanic people crossing over the southern border – is that it's been a non-issue for well over a hundred years. Hispanics would routinely come to the United States for seasonal work in the agricultural sector and then go back to their homes for the rest of the year. People living in Texas and California would routinely cross into Nueva Laredo and Tijuana for leather goods, jewelry, medications and vacations, as well as some "non-family friendly" activities that were harder to find in the United States.

Over the last fifty years or so, there has been several changes in American policy that, inversely, has changed American perception of Hispanics. The first salvo in the modern immigration debate was in 1982, when the Supreme Court ruled on Plyler V. Doe. The case was originally from Texas, and it struck down a law that allowed school districts to deny enrollment to children who were not legally admitted to the United States. From Wikipedia:

"The court majority found that the Texas law was "directed against children, and impose[d] its discriminatory burden on the basis of a legal characteristic over which children can have little control"—namely, the fact of their having been brought illegally into the United States by their parents. The majority also observed that denying the children in question a proper education would likely contribute to "the creation and

perpetuation of a subclass of illiterates within our boundaries, surely adding to the problems and costs of unemployment, welfare, and crime." The majority refused to accept that any substantial state interest would be served by discrimination on this basis, and it struck down the Texas law."

Oddly enough, the next immigration debate centered around immigration legislation that was signed into law in 1986 by a Republican President – Reagan. Reagan had grown up in California and had watched hardworking Hispanics in low wage jobs. The Immigration Reform and Control Act did two major things – it made it illegal for companies to knowingly hire illegal immigrants, and, it gave amnesty to over three million immigrants as long as they could prove they had entered the United States before January 1, 1982, and had resided in the country continuously since that date.

Between these two policy decisions, illegal immigrants now had two major advantages – their children could attend school free of charge, and immigrants couldn't be taken advantage of by companies looking to hire workers at less than a working wage. Immigration into the United States exploded, but more importantly, immigrants weren't going back across the border when the seasonal work was completed for the year – they were staying in the United States because their children could get a better education.

In addition, the Fourteenth Amendment to the Constitution allows for "birthright citizenship" – that is, if a child is born in the United States, even to parents who are illegal immigrants – then that child is automatically an American citizen. This policy was written after the Civil War, and was ratified in 1868. It was meant to apply to former slaves freed after the War, and their children. However, because of the two policy decisions we've already discussed, it gave an added advantage to illegal immigrants. The derogatory term "anchor baby" has been used to describe such a child, in that the parents rely on the citizenship of the child to continue staying in the United States.

With these changes, and the arguable misuse of the 14th amendment, public opinion has changed against Hispanics. To a certain extent it's understandable. Why should a person with no birthright to this country enjoy the same benefits as legal residents? Most undocumented Hispanics and other illegal immigrants are paid either under the table or as 1099 subcontractors – so they don't pay income tax. Why should a group of people who don't pay state or federal taxes benefit from the services that taxpayers do?

The irony here is that another Republican President – Trump – is now using the same arguments against Hispanic immigrants as people did against the Chinese in the late 1800's. "They'll take your jobs." "They're criminals". "They're heathens." It's this type of political posturing and negativity that ignores the real reasons people are frustrated with the immigration issue and invokes an emotional response rather than a reasonable one. Is it a problem? Yes. But it's one that can be fixed in a variety of ways – none of which involve deporting large numbers of people, or holding them in cages hidden away from public sight.

I feel that the carbon footprint of this group of people can be offset in ways that are of benefit to the long-term health of this country, as well as to this group of people and all residents of the North American continent. The people who have made their way to our country illegally do benefit from a "golden door". They are hardworking people that want the best for their families. Use them. These people should be enlisted and trained in green, renewable technology, and put to work rebuilding our infrastructure, weaning us away from fossil fuels, and providing the much-needed talent base that was are so lacking right now. They could be our last, best hope in avoiding the die off predicted in these pages. The fact that our government is so opposed to them, branding them as criminals or as sub-human only shows the kind of ignorance that got us here in the first place.

How We Fund it – Repurposing Existing Funds and Resources to Climate Change

In 2001, just a week after 9/11, Congress passed a resolution that granted the United States President sweeping powers to use force against any nation, organization, or person who aided in the attacks. This resolution, called the Authorization for Use of Military Force, is now being used to conduct military operations in at least seven different countries: Afghanistan, Iraq, Yemen, Somalia, Syria, Niger, and Libya. It has been used to justify military action in other countries as well since the 9/11 attacks – between Presidents Bush and Obama, the AUMF has been invoked for military actions 37 times in at least 14 countries, as well as at sea.

It's been estimated that the wars and military action in Afghanistan, Iraq, Syria, and Pakistan alone has cost the United States $5.9 trillion since the 9/11 attacks. On the other hand, the United Nations Food and Agriculture organization estimated that $30 billion could end world hunger for a year. Not the hungry or homeless in the Middle East – WORLD hunger. I'm not a military expert, but I've got to believe that feeding children makes you a lot more likable than killing them. I'm not a mathematician either, but feeding the world for the last 20 years at $30 billion per year equals $600 billion. We've spent $5.9 trillion in war. Maybe feeding the Middle East and spreading goodwill would have saved us $5.3 trillion dollars.

Of course, that's an exceptionally simplistic approach. On the other hand, consider Occam's Razor: the simplest answer to a problem is usually the correct one.

What if the might of the US military could be used to stop illegal deforestation and mining in South America? What if we used our troops to fight poachers? What if we used our troops to distribute food, medicine, and training to people in central America, rather than spending money on a wall to try and keep them out? What if we used the US military to reforest Africa, South America, and southeast Asia?

The United States spends just shy of $50 billion per year on economic aid and military assistance to other nations. The main US administrative organization than handles worldwide economic aid is the United States Agency for International Development. Their mission statement is "to partner to end extreme poverty and to promote resilient, democratic societies while advancing the security and prosperity of the United States." They work in low-income, developing countries on a long-term timeline to prevent poverty, develop a country's socioeconomic base, and provide technical cooperation on global issues, including the environment. Don't get me wrong – I'm not arguing against this money. It's less than 1% of the US government's annual budget.

The US spends $100 billion per year in corporate subsidies. This includes farm subsidies, rural subsidies, SBA loans, as well as $4 billion a year for energy subsidies. Granted, a portion of these are supposed to be used for R&D for alternative power sources. The percentage of money given to renewable energy sources is historically low compared with subsidies for fossil fuels, but it's still there. However, Shell is one of the top receivers of government subsidies, and they are one of the most profitable companies in the history of the world. Shouldn't Shell support it's own R&D into alternative power sources?

The US still spends $1 billion per year on the War on Drugs. While most Americans associate the War on Drugs with Reagan, it was actually Nixon who began the initiatives of "prevention of new addicts, and the rehabilitation of those who are addicted" in a press conference in June 18, 1971. The War on Drugs has cost the American public an estimated $1 trillion since 1971.

Our military has been used in black ops all over South America and countries north and south of the Panama Canal to try combat drug lords and drug processing areas. In 2015, the federal government spent an estimated $9.2 million EVERY DAY to incarcerate people charged with drug-related offenses, regardless of the fact that in June 2011, the Global Commission on Drug Policy released a critical report on the War on Drugs, declaring: "The global war on drugs has failed, with devastating consequences for individuals and societies around the world."

The real mother lode, however, is the US military. The United States spends more than any other country in the world on it's military - about $650 billion per year on military spending. That's more than the next ten countries military spending combined, all of which are our allies. China has the second highest military expenditure in the world, and they only spend $250 billion annually. That's 38.5% of US military spending.

I don't think we should cut military spending. I don't even think we need to change the mission of our military. From the Military.com website: "The United States Military branches exist to defend the United States against all enemies and to provide combat capabilities anywhere in the world in support of United States security objectives." This is a laudable goal and I have nothing but the highest respect for the members of our Armed Forces. The US military is the most technologically advanced fighting force in the history of the world. Here's my question – why can't we use this resource, even in part, to save the world from climate change?

What is Stopping Us from Using Presidential Power to Enact a New Conservation Corps?

The President has both formal and informal powers. The formal powers of the Presidency come from Article 2 of the Constitution, and they grant the President the ability to do a number of things. The President is the Commander in Chief of the Armed Forces. He or She can choose ambassadors and choose appointees to the Supreme Court (with Senate approval). He or She can appoint members of their own cabinet. He or She has the ability to negotiate trade deals with foreign countries. Remember that - we'll get back to this in a second.

Informally, the President has other powers, and they have been growing since around Roosevelt. The most used informal power that the President can command is the Executive Order. Several of the most famous turning points in the history of the United States were Executive Orders: Lincoln's Emancipation Proclamation was an Executive Order, and ironically, so was Roosevelt's internment of Japanese-American citizens during WWII.

Another informal power the President has is to discuss and negotiate agreements with heads of foreign governments. If you add the powers that the Authorization for Use of Military Force gives the President we discussed in the previous chapter, then the informal powers granted to the President provides almost unilateral power in discussing trade, use of force, and other agreements with foreign nations. Our current President has used these powers (some say mis-used) by waging an inflammatory trade war with China, by forcing the Mexican government to detain refugees on it's southern border, and by visiting both allies and hostile nations on their own soil.

The President also has the power to assign and built new branches to the United States Military. On December 20, 2019, Trump signed into law the US Space Force, the sixth independent branch of the US Military, and the first to be signed into law in 70 years.

So – to answer the question – What is stopping us from creating and implementing a climate conservation corps that will benefit the North American continent, and the world? The answer:

The signature of one person. Anyone occupying the office of the Presidency could sign this organization into being. They have the power to negotiate an agreement with Mexico and Canada, the power to create such an organization by use of Executive Order, and the power to command the US military to support it.

Mr or Ms President – I would like to take this opportunity to formally and unequivocally state that climate change is a Clear and Present Danger to the United States, and in fact, to the world. I would formally offer my services to combat this threat, and become a member of a conservation corps specifically created to address and negate the tide of destruction we will surely face by ignoring the evidence. If I can assist in any way, I am able and willing to follow your instruction and command as a citizen patriot of the United States of America.

To make a climate conservation corps work, the mindset of our leaders has to change. The mindset of our policymakers, lawmakers, and the mindset of the President has to change. I fear that the only thing that will change the mindset of our leaders IS the mass die off I've predicted in these pages.

Maybe there is still hope.

Chapter 6: The Tipping Point – How it Begins

Earlier in this book, we talked about Stalin's Great Famine, and Mao's Great Leap Forward. I like to call these 'combination events' because they were caused not just by one factor, but a variety of factors that contributed to a massive death toll. In Stalin's case, the famine was pre-meditated and directed at the Ukrainian public, as well as adding other factors of political exports and natural disasters that affected the harvest. Unlike Stalin, Mao didn't intentionally kill off his countrymen – but the combination of bad political policies, exports, and crop overcrowding still killed 43 million people.

Most historic tragedies were combination events. They've been called a 'perfect storm' of events – some might even call them a 'comedy of errors'. For example, the sinking of the Titanic wasn't attributable to a single factor – the ships design, the inability to turn, even the fact that the watchers in the crow's nest didn't have binoculars on the night of the tragic accident all contributed to the disaster.

The beginning of the modern human die off will be another combination event. Each continent will experience their own version of this 'tipping point', and I believe that they will happen in rapid succession of each other. After all, WWII began in Europe and within a space of a few years pulled the entire world into calamity. Why should the next die off be any different?

If the tipping point is caused by climate change (as most scientists think it will be) we will begin to see the effects at around 2 degrees of warming. There will still be a tipping point – one large event, or a series of events that will definitively be where the die off begins. In each continent, the site of the event will become as famous as Auschwitz, the 6th Floor in Dallas, the Lincoln Memorial, the Killing Fields in Cambodia, or any other area where history stood for a brief amount of time.

I don't know which continent the first tipping point occur on, but I can make some educated guesses. The entire continent of Africa is going to be the most affected by climate change, and heat exhaustion is going to become a major factor in deaths there. Any country south of Turkey and north of the Sahara is also in imminent danger, due to civil strife in those area that will eventually cause a complete collapse of infrastructure. A three to five degree rise in temperature in India will cause the entire country to boil, and send climate refugees fleeing north.

I think the Americas will fare better than some, but we are still going to see millions of deaths in northern South America, Panama, Mexico, and into the United States. Depending on how Canada handles the influx of climate refugees, they will handle the die off much better than their neighbors to the south. Looking to Europe - Scotland, Finland, Ireland will all avoid the high numbers of deaths their southern neighbors will see – again, depending on how well these countries can secure their borders.

In Russia, the largely landlocked countries of Siberia, Belarus, and the Ukraine will also do well. Their survival, and the survival of areas in Southern Europe, will depend on the way other countries secure their borders. For example, if Morocco cannot or will not secure the flow of climate refugees from northern Africa, you'll likely be able to walk across the Strait of Gibraltar into Spain on a blanket of floating corpses. Likewise for Turkey and their neighbors that border the Black Sea.

The scenarios below are based largely around events I expect to happen in North America. Some are based on climate change, and some are based on an outside aggressor. Some are die offs created in part by natural disasters, and some are based on political events. However, I don't believe that any of these are fiction, or I wouldn't have put them here.

I'm not a fortune teller. However, I really don't need to be one. All I have to do is listen to the scientists and pay attention to the lessons of history. With that thought held firmly in mind, here are three possible scenarios (with, of course, historical backing) that will start the next human die off.

Scenario #1: The Perfect Storm of Rising Seas, Increasing Intensity of Storms, and our Crippling Lack of Immigration Reform

Although most people don't realize it, the deadliest storm to ever hit the United States wasn't Sandy, or Andrew, or even Katrina. In fact, it's the deadliest natural disaster to ever hit the United States to date, and it doesn't even have a name.

On September 8, 1900, a category 4 hurricane struck Galveston, Texas. This was another combination event that was exacerbated by many factors. The topography of Galveston Island is that of a large sandbar off the Texas coast – the highest point of the island is only 5 feet over sea level. Galveston had endured many storms during its development, and in the spring and summer of 1900, was a bustling, tourist boom town that saw ships and guests from all over the world. Residents who had weathered several storms had discussed building a sea wall, but it was dismissed as unnecessary by city government. In fact, the Galveston Weather Bureau section director Isaac Cline wrote an article in an 1881 edition of the *Galveston Daily News* that it would be impossible for a hurricane of significant strength to strike the island. Another factor in Galveston's imminent destruction was that local builders would take sand from the beaches to fill in loose holes or areas of construction – further exposing the island to rough seas.

In 1900, the most advanced meteorological station in the world was in Cuba. In the days preceding the Galveston storm, the director of the Belen College Observatory in Havana tried to communicate with the United States Weather Bureau (the predecessor to the National Weather Service) to tell them that the storm was on a western path through the Gulf of Mexico. However, the advice was ignored because of recent political tensions from the Spanish American War. It wasn't until September 6 that a ship's captain named TP Halsey reported the storm at 100 mph windspeeds in the Gulf. Even armed with this information the Service would not call the storm for what it was – at the time, it was protocol to not use the terms "hurricane" or "tornado" in weather alerts, in order to keep from panicking the population.

On the morning of September 8, the day dawned partly cloudy, although larger than normal swells were reported. By now, US and Galveston weather officials knew the trajectory of the storm, but it was too late to warn the majority of the population. The ones that did receive warnings were unconcerned by the clouds that began to roll in from the Gulf around noon, and very few people used the bridges to evacuate to the mainland. By 5 pm, reports of 100 mph winds were recorded in Galveston. Although the residents couldn't know it because of the cloud cover, the sun set on Galveston at 6:33 pm. The storm officially made landfall in Galveston at 8 pm. Many residents wouldn't live to see the morning.

Words on paper can't convey the fear that these people would have felt. At 140 miles per hour, anything becomes a flying projectile – from bricks and stones, to trees, cars, and houses. The storm surge that struck Galveston washed over the entire island. It was 12' high, and the island only sits at 5' over sea level. Buildings that were built to resist hurricane winds were damaged by other buildings that had been knocked off their foundations by the storm surge. Not one single building in Galveston was undamaged.

Imagine for just a moment listening to the wind scream while the water rises to your front door, and into your house. Imagine the roof being blown off, and you being thrown to the floor while the entire house is shifting off it's foundations. Imagine your children screaming as the wind and wind-driven debris literally breaks the walls of your house down. Now – imagine it all happening *in the dark*.

The final death toll will never be known. Official records put it at 8,000, but most reports are anywhere between 6,000 and 12,000 people. Because the bridges to the mainland were all destroyed, Galveston was isolated in a way that New Orleans, nearly 100 years later, wasn't. Initial reports of aid workers described screaming from the rubble of the buildings the first few days after the storm, but there wasn't anything that anyone could do – there weren't enough survivors left to save people trapped in what was left of the buildings. After a few days, there wasn't any more screaming.

When Hurricane Andrew hit Florida in 1992, many of the factors that made the Galveston storm so devastating had changed. Today, radar shows us the path of a hurricane long before it makes landfall, giving the residents time to evacuate. Nearly 1.2 million residents in Miami-Dade country, and in the upper and middle Keys evacuated before the storm made landfall at Homestead, Florida on August 16, 1992. Those that couldn't evacuate were ordered to storm shelters specifically designed to withstand hurricane force winds. The storm surge was reported at just over 16' at Homestead, and, at the time, the Category 5 storm caused more financial damage than any other storm in Florida's history. Most damage was caused by the 175 mph wind speeds generated by the storm. Flooding was minimal, and with the exception of Homestead, most of the rest of the coast only saw 4'-6' storm surges.

I don't mean to lighten the trauma of Andrew – it was a devastating hurricane that forever changed people's lives. In the storm, and the aftermath, 44 people lost their lives, and thousands were left homeless. However, Homestead isn't the most vulnerable area in the State of Florida. That distinction arguably goes to South Beach, Miami, 45 miles to the north.

Since 1992, sea level rise due to climate change is just a little over three inches. Although historically we have seen the seas rise about 1/8 of an inch each year since 1992, sea rise will increase and accelerate in the coming years due to the acceleration of climate change. By 2100, experts at NOAA estimate that sea level rise will increase by 6 feet. Miami is only 6.5' above sea level. And South Beach doesn't have a continuous seawall.

It has been said by climatologist that there is no scenario that will see Miami above water by the end of the century. But I believe that the tipping point will come much sooner than that. I also believe that at least one scenario of the tipping point to the coming die off starts in Miami.

The eye of a hurricane can be anywhere between 15 and 40 miles wide. The barrier island that protects Miami is an 11 mile stretch of shoreland from South Beach to North Miami beach. If South Beach took a direct hit from a Category 5 tomorrow, it could cause a storm surge that would overtake the barrier island and push into Biscayne Bay. However, if it happens in 20 years, with the level of ocean rise significantly higher than it is now...

Katrina pushed it's storm surge 12 miles into the Mississippi River. A storm surge with a drastically elevated sea level change would hit Miami with nothing less than the force of a low yield nuclear warhead. Even with South Beach and Miami Beach acting as a barrier island, a 16' to 24' tall wall of water would push through South Beach, through the Bay, and push inland from Downtown Miami through to the Upper East Side. This wave of water would continue inland to Hialeah, Fontainebleau, and Palm Springs, and push up to Interstate 75 and beyond, which would effectively cut off Southern Miami, and the Keys.

Katrina lasted for 6 hours at Hurricane force winds in 2005. However, Dorian hammered the Bahamas for more than 24 hours at Hurricane force winds in 2019. We know that Andrew caused the evacuation of 1.2 million people in 1992. We know that storms are becoming more powerful because of climate change.

The population affected by a storm directly hitting South Beach today would be well over 3 million people. If the storm stalls like Dorian did over the Bahamas (which is only 130 miles away from Miami) then we are looking at a scenario where these people would simply have nothing to return to. No city on Earth can withstand 175 mph winds for 24 hours. Furthermore, there isn't a city in the United States that could handle an influx of that many people so rapidly. The closest major cities are Tampa, Jacksonville, and then, Atlanta, but none of them has a population of over a million people. Tampa has less than 400,000.

The only way that the situation could be dealt with is large refugee camps in the northern part of Florida and into Georgia. Here's the bigger problem. South Florida is home to one of the largest populations of undocumented immigrants in the United States – it's estimated that nearly half of million people living in South Florida are illegal immigrants. It's the fifth largest undocumented population in the United States. Imagine a situation where half a million undocumented immigrants are placed in government sponsored refugee camps. How long do you think it would take for the deportations to begin? How quickly would they turn violent?

Food and water supplies would become critical very quickly. Even with the support of the Red Cross, 3 million displaced climate refugees with nowhere to go home to would stretch emergency resources to the maximum limit and beyond. Remember that Katrina displaced half of the population of New Orleans, about 250,000 people. Remember that it took FEMA 5 days to get fresh water to the Superdome. How long would it take the government to resettle 3 million people?

There comes a point where fresh water, food, and septic becomes more than what the affected area can supply. History shows us that the steps leading up to this critical point include profiteering and violence towards the affected people. History also shows us that the response to groups of refugees is overwhelmingly negative. As food supplies run out, the people that initially begin to suffer the effects are the refugees. The affected area begins to spiral into an ecological collapse, as it is inevitably overharvested. Soon, the local residents begin to be affected. Starvation deaths begin, and increase in number. Violent acts against refugees begin, and increase in number. Finally, you end up with armed camp where people are fenced in, with military guards to make sure no one leaves. That's when the dying begins in earnest. Is it possible for the United States, one of the most prosperous developed nations in world, to have a refugee camp full of starving climate refugees within it's borders?

Remember, however, that the tipping point will be a combination event. While the prospect of a Category 5 hurricane hitting Miami is frightening, I don't think it will be enough to begin the die off I've predicted in these pages. There would have to be another factor that happens at the same time to truly start the spiral.

There are two other factors that I believe are tinderboxes waiting for a spark in the United States, and both of them are completely overlooked by most Americans. These issues will work in combination with sea level rise to begin the die off.

The first is the homeless problem in America. California leads the nation in the homeless population. Los Angeles has somewhere around 60,000 homeless alone, out of 130,000 in the state of California. These are official numbers – which is almost certainly an underestimate, as homeless people can and do couch surf with friends or family, stay in hotels when they have the ability to do so, or sleep in cars. No one really knows what the number of homeless are in California, or the United States as a whole.

That doesn't seem like many people, but the homeless are susceptible to a much higher level of disease than the rest of the population. People that are living on the street are at much higher risk to everything from the common cold to HIV. Tuberculosis is a real problem, as it is a disease that requires a very long treatment cycle, and homeless people are at risk for not completing treatment when infected. Homeless people have a higher incidence of substance abuse, which can be a transmission vector for TB. Right now, it's estimated the 5% of the homeless population in California has an active TB infection.

The second factor in the tipping point could well be a disease that uses the homeless population as an infection vector. I don't think we are likely to see a smallpox or polio epidemic – smallpox has been eradicated worldwide, and polio nearly so – but local health experts in California have named typhus, leprosy, and our old friend plague as possible epidemics among the homeless in California in just the last few years.

A plague epidemic in Los Angeles would be bad, but I don't think it would empty the city. I do think that, in combination with another major event in the United States, it would cause a level of civil unrest that might be pervasive enough to affect the entire country. I also think it would stretch the resources of healthcare professionals and aid workers in the United States. Plague in the modern age is treatable. But who would go to the tent cities and treat it? More importantly, would a group of people who have been outside normal medical care for so long accept treatment?

The third factor in this scenario involves the southern border of the United States and the immigration problem. This issue has been neatly swept under the rug – for now. But for how long? I believe that the temporary band-aid enacted by the Trump administration by forcing Mexico to deal with immigration into the United States has made a long-term problem into a potential powder keg, and most Americans are blissfully unaware of it.

For years, the Cartels have ruled much of Mexico. Beheadings, mass graves, retribution and revenge attacks have all seeped into American news. We've been witness to dismemberments in tourist areas, and the murders of elected officials, police, and journalists who have gotten in the way of the drug gangs, or said too much, or wouldn't take a bribe. Mexico has one the most corrupt police forces in the world, and bribery is common. Pablo Escobar said it best: "Plata O Plomo"; silver or lead. Meaning – take the silver (a bribe) or take the lead (a bullet).

But recently, more ominous rumblings have been coming from our neighbor in the South. In January of 2019, a pipeline carrying gasoline exploded in Hidalgo country. Over 100 people burned to death in the explosion, which was said to be caused by gasoline thieves drilling into the pipeline. This wasn't done by drug gangs – it was done by desperate people trying to feed their families.

Another incident, involving the townspeople and their elected mayor in Chiapas State in southern Mexico occurred in

September of 2019. The townspeople, angry that the mayor had not kept his campaign promises to fix the local roads, forcefully dragged him from his office, tied him to a truck bed, and dragged him through the streets of the town. It took dozens of state police to stop the truck and save the Mayor.

Most troubling is the rise of pirates in the Gulf that are preying on tourist boats and oil drilling rigs. Pirates dressed as police have boarded tourist launches and stolen everything from the groups of tourists on board at gunpoint. Oil drilling rigs operated by Pemex are at the most risk, with some platforms being boarded by pirates who lock up the crew of the rig at gunpoint while stealing equipment and supplies.

This isn't a small issue, although we don't hear much about it in the United States. Gasoline and oil theft have been reported across Mexico at over 1 million barrels per day, with an average of $1 billion annually in losses. Theft of oil, and oil related tools and hardware are up 318% in 2019 over the last two years. This money is going to support criminal organizations with the tacit approval of the Mexican people, who can buy fuel on the black market cheaper than they can from a state-owned gas station.

Additionally, the legalization of marijuana has cut a huge chunk out of the profit margin of the drug cartels. Although the meth, heroin, and opioid trade is still building fortunes in the hundreds of millions of dollars to the Cartels, the loss of income due to legal marijuana has hurt, and the Cartels are turning to other forms of income.

Now, we have the issue of immigrants being detained inside Mexico instead of travelling overland into the United States. We've already discussed how these immigrants and refugees from Central America and the northern countries of South America will be treated by Mexican nationals. Xenophobia and jealousy will abound as these immigrants utilize resources that the resident population is already in short supply of. In turn, these immigrants will become targets of police and gangs looking to extort any amount of remaining money or resources these people have.

So, you have a poor population made poorer by corrupt police and Cartels. We are starting to see nascent areas where the population is beginning to rise up and depose their elected officials violently. We are starting to see armed groups of pirates attacking government owned facilities. What is the final outcome to this recipe? In a word – revolution.

The United States has a long history of meddling in the governments of Latin America. Frankly, I don't believe that the United States would allow a vocal, outspoken opposition leader in Mexico to continue breathing. However, imagine for a moment a modern-day Che Guevara, or a Fidel Castro, or an Evita Peron in Mexico. It's one thing for the United States to screw with Venezuela or Columbia – they're on the other side of the Panama Canal. But an opposition party in Mexico? Led by someone who had a nation of 129 million pissed off people behind them?

There may be people who accuse me of being naïve, but I think the possibility is a real one, especially considering two things – the rise of sea level in the Gulf and the Pacific, and the loss of farmland in Mexico due to climate change. Now, you have 129 million *starving* pissed off people looking to their fat and happy neighbor in the north.

Combine this with the mindset of the American political population, and we may end up with a real problem in 20 years or so. Imagine an opposition leader in Mexico amassing a crowd of refugees at the southern border of the United States. The smart leader would pick a smaller border town – probably Eagle Pass. You wouldn't even need an armed force, which is just as well, since Mexico has extremely strict gun laws. If our hypothetical opposition leader could get, say, 3 million people ready to storm the border and begin a peaceful entry into the US, and have the organizational skills necessary to have busses, water, food, and people ready to accept the immigrants, you'd have a real political mess on your hands.

At some point, the numbers of people guarding the fences cease to matter when faced with a significant number of people trying to get in. It's a simple numbers game. The only way to win it is to start firing on people. Would the people of the United States condone the mass killing of peaceful refugees on our Southern border? How would the United States public react to armed border patrol agents and national guard members indiscriminately firing into a crowd of political and climate refugees full of women and children?

Even if our hypothetical opposition leader used weapons to storm the initial crossing, a fight of say, 1500 border agents, or 5000 national guardsmen are no match for the amount of people that would stream into the United States if the border was seriously compromised.

This is the third and final vector in this first scenario. How would our government react to this type of combination event – a superstorm in South Florida creating 3 million homeless climate refuges, a contagious disease outbreak affecting the west coast, and a compromised southern border letting in millions of illegal immigrants?

You'd see a complete collapse in aid. Schools, medical care facilities, government health programs – all would collapse under the weight of that many people needing help. More importantly, the clean water and food resources in the southern United Sates from New Mexico to South Carolina would be severely impacted by this many people.

The final straw would be the loss of government aid to the 42 million people currently on SNAP (Supplemental Nutrition

Assistance Program, otherwise known as food stamps) in the United States. Rising food prices due to shortages in the South would have a dramatic impact on the amount of food people on government assistance could buy. The massive influx of displaced people and climate refugee would wipe out nonprofit food banks across Texas, Louisiana, Mississippi, and Georgia. The first deaths from malnutrition would begin in the very young and the very old. Elderly on fixed incomes and no method of transportation would starve to death in their homes. You'd start to see more and more instances of civil disobedience. The tipping point has come. The government and ultra-wealthy wall themselves into their fortified, gated communities, and abandons the rest of us to survive, or die.

Scenario #2: An Outside Aggressor, Ammonium Nitrate, Social Media, and Deepfake Technology

The threat of nuclear war has been a shadow on the world since before most of us were born. However, there are a lot of ways to seriously cripple the infrastructure of any developed country that doesn't involve a nuclear weapon. Its relatively easy and inexpensive to use the technology of a developed nation against it's own people. Using these tactics of reverse technology weakens the economy of a country, and demoralizes the population. These types of tactics could be used to set in motion a chain of events that would cause the tipping point we've discussed.

Here's a date to remember – April 16, 1947. This is the date of the worst industrial accident that ever occurred on American soil, and it happened at Texas City, near the Texas/Louisiana state line. The *SS Grandcamp* was a French ship carrying ammonium nitrate produced during WWII, and was now destined to be used as fertilizer. The ship was carrying 2300 tons of the chemical, and when smoke was noticed in the cargo bay at around 8 am that morning, the crew (presumably) made the decision to extinguish the fire by smothering the flames, rather than using water (which would have ruined the cargo).

At 9:12 am, a small portion of the chemical reached 410 degrees. It triggered an explosion that vaporized the ship itself, destroyed the entire dock, and created a blast wave that flattened homes and businesses. The burning shrapnel ignited oil tankers that burned for days, and completely destroyed a nearby Monsanto chemical processing plant, oil refineries, granaries, and most of the warehouse district.

The blast was heard 150 miles away, and it knocked people off their feet 10 miles away. Near the epicenter of the blast, 600 people were killed, including the entire Texas City fire department and all their equipment, which severely hampered efforts to contain the fires. Thousands of people were injured. The explosion created a 2000 foot tall mushroom cloud, and it also triggered a 15' tidal wave, further damaging the docks after the initial blast wave.

"Well," I can hear the reader saying, "that happened in 1947. Surely the government monitors dangerous chemicals like that far better now." Except – they don't. In 2013, in a small town called West, Texas, was devastated by a chemical explosion. This fertilizer plant had a history of petty thefts of it's ammonium stores, presumably by employees that were using it to make methamphetamine. It had also written violations for improper storage of chemicals and safety violations with multiple government agencies dating back several years.

In the evening of April 17, 2013, fire broke out at the plant, and the fire department was called in to fight the fire. Twenty minutes after the fire broke out, first responders had already arrived on the scene and were attempting to put the flames out. They had no idea of the amount of chemical stored on the site, and 240 tons of ammonium nitrate exploded, killing all the first responders and creating a 95 foot crater where the chemical plant had once stood. Fifteen people were killed in the explosion – which was later determined to have been a result of premeditated arson.

It's not just a United States phenomenon, either. On August 12, 2015, a series of explosions killed 173 people at a container storage facility at the Port of Tianjin, in China. In this case, a container with improperly stored nitrocellulose (also called guncotton) exploded, triggering a much larger explosion about 40 minutes later of a container filled with 800 tons of ammonium nitrate nearby. There was also 700 tons of sodium cyanide stored at this facility, which, after being spread by the explosions, caused a bright white chemical foam to cover the streets of the provinces during rainfalls on August 18 and 25.

We'll never know what other chemical were stored at this processing facility, because a) it's China, and b) most likely, the Chinese government probably doesn't know either. We do know that the blast registered as a magnitude 2.9 earthquake, spread fire and ash thousands of feet into the sky, and was the equivalent of nearly 22 tons of TNT. That's the same explosive force as a low-yield nuclear weapon.

Ammonium nitrate is cheap. You can buy it in almost any farm supply store in the United States. In fact, the bomb that Timothy McVeigh constructed when he and Terry Nichols bombed the Federal Building in Oklahoma City was made primarily of ammonium nitrate bought in 50 pound bags from rural agricultural supply stores. The total cost was about $5000. When that bomb exploded on April 19, 1995, it killed at least 168 people and damaged buildings in a 16 block radius.

Ammonium nitrate is imported and exported all over the world. Unsurprisingly, Brazil is the largest importer, although the United States also imports a hefty amount. Russia is the world leader in ammonium nitrate exports.   Almost all import and export is done by sea. There are some countries that have banned or specifically curtailed ammonium nitrate, or enacted strict security measures on it, but they are a rarity.

Let's add another factor to this scenario. Social media now affects the decision making process of every consumer in a developed or developing country on the planet. In fact, it's becoming more rare that social media is NOT a factor in decisions we make in our everyday lives. According to one survey, 74% of consumers make a choice on product purchases based on social media. It affects where we shop, who we meet, what we eat and drink, where we live, educational choices, job choices – and who we vote for.

United States intelligence knows for a fact that there was Russian meddling in the 2016 presidential election. Thousands of false accounts were created in a Russian-led think tank in St Petersburg specifically designed to promote Trump and discredit Hillary Clinton. Hackers associated with the GRU (the Russian version of the CIA) infiltrated information systems of the Democratic National Committee and other Democratic organizations, and then leaked documents through Wikileaks and other platforms.

This wasn't the first time that Russia used social media to further their intentions. The May 2014 Ukrainian elections were severely disrupted by cyberattacks, including hacked emails, disinformation campaigns, denial of service attacks, and attempted alteration of vote tallies. In fact, some of the same malware used in the Ukrainian hack was found two years later on the servers of the Democratic National Committee, providing a concrete link between the cyberattacks in the Ukraine in 2014 and the Presidential election in 2016.

After 9/11, the United States recoiled in shock and horror from the terrorist acts. Civilian airspace was completely shut down in the US and Canada for three days after the attacks. The US Stock Market didn't reopen for six days. All across the United

States, events were cancelled, schools were closed, and sporting events were postponed or cancelled. Any large gathering of people was met with fear, and the winter shopping season in malls and department stores was especially subdued. Arguably, part of the fear associated with the terrorist attacks of 9/11 is that we, a nation, didn't definitively know for sure who was even responsible for the attacks. Al-Qaeda and it's leader, Osama bin Laden, didn't take public responsibility for the attacks until 2004.

The internet was a tool used primarily by business, government, and educational facilities in 2001. Most people in the developed world didn't have internet at home, and the ones who did had dial up. Social media hadn't been invented yet. Facebook was just a twinkle in Mark Zuckerberg's eye – it was started in February of 2004. Google was the equivalent of a toddler at three years old. Video wasn't available on the internet. Credit card transactions online were met with suspicion. The Nokia 3210 was one of the most popular cellular phones in the world. It had no GPS, no Bluetooth, and no internet browser.

It may sound like a Tom Clancy novel, but imagine for a second: a foreign power or terrorist group that wanted to harm the United States, or any developed nation. It's counterproductive to launch nuclear weapons. They're expensive and pretty much guaranteed to invoke a very nasty response. They're trackable – US weapons detection systems can pinpoint a launch site within 5 feet, anywhere in the world – and have been able to for at least the last 20 years. Why not use a combination of a powerful but cheap explosive, and capitalize on it with social media?

So, here's the basis of scenario #2. A government or group who doesn't want to be identified decides to destabilize the American people by simply floating a tanker or carrier with a payload of ammonium nitrate into a major United States port, or a port of one of our allies. You can pick the country, as there are several that are unhappy with the United States foreign policy right now. China has been the target of US led trade wars, and is not happy with our involvement in Hong Kong. The US has meddled in Latin American affairs for decades. The US withdrawal from Turkey has both sides unhappy, and we actually store nuclear weapons in Turkey. The killing of Iranian General Soleimani. The chessmasters in the Russian government that have deliberately and intentionally manipulated American politics. And, of course, most of the Middle East.

An intentional detonation of the size that happened in Texas City in any American port would be devastating. Using a shaped charge or fragmentation materials would cover a much larger area than the accident in 1947, or the explosion in West, and significantly increase the body count and damage of the blast. However, the real key to our scenario is the use of social media.

A targeted social media campaign in conjunction with an uncredited explosion would incite fear in a way nothing in the developed world has ever seen before. Imagine it: a country or a terrorist group detonates a 500 tons of ammonium nitrate mixed with fuel oil – commonly called ANFO – and then launched a completely untraceable social media campaign telling Americans or American allies that the next explosion will occur in another major port in a week. In addition to the tragedy, destruction of property, and loss of life from the first bomb, the entire population sees trending news reports and social media posts of another bomb being set.

The technology that would be most valuable to our hypothetical aggressors is called "deepfake", and it involves the manipulation of video. This manipulation typically involves putting someone else's face on a public figure, and appearing to be that person. As an example, our aggressors detonate a bomb in a major port. Less than 30 minutes later, a video with President of the United States appears on social media that shows him or her telling the

American people that their government has received information that the bomb was a nuclear explosion and that the surrounding 250 miles of the port city is susceptible to nuclear fallout, and to use duct tape to seal your doors and windows. Furthermore, that the US has intelligence that the next weapon is set to go off in another major American city in less than 24 hours.

People have been trained to believe social media. And with a deepfake video, the term "seeing is believing" takes on a new, horrifying reality. An ANFO bomb would create a huge mushroom cloud – the one in Texas City was 2000 feet tall. People would take video of it and post it to their own feeds. Our aggressors wouldn't even have to set off a second detonation. Fake video of a detonation in any major city, or several major cities, would be instantly believable based on the news reports coming from the first detonation. Fake video of our President saying that a major city had been completely destroyed would only fan the fire.

What's worse is that even if our government got online and on television and disputed a deep fake video, there would be a large number of the population that would believe the fake videos over the real ones. As I've said before, the credulity of the American population is at it's limit. People know that their government lies to them on a constant basis about a variety of issues. People know that government isn't for the people anymore, it's a way of keeping the poor and middle class quiet while the wealthy and ruling class take all they can. Media and our government has already blurred the lines between what is real news and what is "fake news".

The consequences of such an attack would be completely catastrophic. You'd see a mass exodus from every major city in the United States. FEMA would issue a "shelter in place" directive, but most people would disregard it, believing that the government was trying to avoid a panic. You can't enforce a curfew in every major American city – you'd have to deploy the National Guard to maintain a curfew, and it would be impossible to enforce across the entire nation.

Highways would be completely and utterly impassible. People would run out of fuel in traffic jams hundreds of miles long and die of dehydration literally parked in the middle of an interstate highway. Small towns would barricade freeway exits and place armed guards to keep a literal hoard of starving and dehydrated refugees out of their communities. The population shift of an entire city within a day would cause massive disruptions in food delivery, sanitation, fresh water, and communications. Profiteering would be common. Would you pay $100 for a bottle of water if your child was dying of dehydration in the back seat of your car? $500? $1000?

Let's say the attack took place in August. With climate change, you'd be looking at hundred degree temperatures across the South. Most people simply aren't used to being outside in those types of temperatures for more than a few minutes. Heatstroke would cause thousands of deaths in the first day, and millions in the second day. Dehydration and starvation would soon follow. The elderly and the very young would be the first to die, but within a few days, there would be more dead people than living people.

When Hurricane Rita threated Houston in 2005, the wounds from Katrina were still raw. Houston ordered a mandatory evacuation, and 6.5 million people flooded Texas highways before the storm made landfall on September 24, 2005. People sat in traffic for over 24 hours in 100 degree weather. Dozens died on the highway from heatstroke – while back in Houston, only 10 people were killed from the hurricane itself. The evacuees from Rita were travelling north. But what would happen if there was no place to go? An entire nation of people on the roads, blindly fleeing what they thought was real danger?

How long could something like this last? The true frightening answer is "indefinitely". Once people began to calm down and return to their homes, our aggressor would just release another round of videos. But let's say that the crisis lasted only 10 days. Could you survive 10 days with your family, locked in a car, with no water, no food, and no gasoline? Even if you decided to go back home, you couldn't – the traffic jam would keep you in place.

As we've discussed, the population of the United States is 327 million people. An attack like this might cause a loss of 10% of the population in less than a week. Within weeks or days, our aggressors have selected another target in the developed world, and the cycle repeats itself.

"Ha!" I can hear the reader saying. "There has only been one foreign country that attacked United States soil, in the last 250 years, and look what happened to them!" And that's undoubtably true. Japan suffered greatly after the attack on Pearl Harbor. The bombs dropped on the (ahem) military targets of Hiroshima and Nagasaki proved beyond the shadow of a doubt that the atomic age of warfare had arrived.

However, there's a big difference between "haven't yet" and "will never". In addition to our port cities, the infrastructure of the United States – our freshwater processing, water reservoirs, and dam systems; our power transmission lines and substations; our bridges, roads, and transportation systems – these are all huge targets for both physical attacks as well as digital attacks. Now, in the age of drone warfare, you don't even need personnel on the ground to coordinate an attack on these types of facilities. These targets, along with our port cities, could be the tipping point for the next die off.

Scenario #3: Wildfires, Food Production Pathogens, and Opening of Pandora's Box in the Arctic

Most Americans will remember learning about the Great Chicago Fire of 1871 in school. Through a combination of events, this fire destroyed nearly 1/3 of the city, and an estimated 300 people died. This fire has been attributed to Ms. O'Leary's cow knocking over a lantern as it was being milked (the cow, not the lantern). While it's true that the fire began in the O'Leary barn, the truth of the origin of the fire will never be known. Interestingly, the O'Leary's were targets of both anti-Irish and anti-Catholic sentiment, as the family were Irish immigrants. Although they were never officially charged with starting the fire, the Chicago City Council exonerated the O'Leary family – and the cow – in 1997.

What makes the Chicago fire so memorable was arguably not the fire itself, but the outpouring of support from around the world to rebuild Chicago. In fact, a donation from the United Kingdom was used to create the Chicago Public Library after the fire – one of the first free public libraries in the United States.

In an unusual foreshadowing of today's environmental crisis, a town near Chicago was completely deforested supplying timber to rebuild Chicago after the fire. The town of Singapore, Michigan, cut down all it's trees to supply the lumber, and as a result, was overrun by sand dunes from the wind and water coming off Lake Michigan. The town was literally engulfed in the growing dunes, and had to be abandoned in 1875.

One of the conditions that caused the Chicago fire to be so devastating was a very dry summer in 1871. This led to other fires in the region, most notably the largely forgotten Peshtigo Fire. This wildfire was the deadliest wildfire in American history, and it occurred about 250 miles north of Chicago. The Peshtigo fire was much larger than the Chicago fire, with a greater loss of life – while the Chicago Fire destroyed about 4 square miles, the Peshtigo fire destroyed 1850 square miles and killed somewhere between 1200 and 2500 people. The true number will never be known because, out of the 12 small towns in the path of the fire, all were completely burned to the ground, including all the local records.

One of the reasons the Peshtigo fire is largely unknown, even though it occurred on the exact same day as the Chicago Fire, was the complete lack of communication coming from the fire. The telegraph lines coming from Peshtigo were all destroyed in the beginning of the fire, leaving it to burn almost completely unchecked.

The Peshtigo fire is notable for many reasons. It literally incinerated entire families. It caused a huge loss of life and a significant area of land to be completely destroyed. It was also the first documented firestorm in the United States.

A firestorm is more than a wildfire. A firestorm is a large enough fire that it creates it's own atmospheric conditions around it that support and exacerbate the fire itself. As the fire grows in intensity, it requires oxygen to burn. A normal wildfire is driven by the wind, but a firestorm develops it's own winds from the consumption of oxygen from around the entire edge of the fire, in all directions. This wind can reach speeds up to 110 miles an hour, can draw items into itself (such as cars, animals, trees, and people), and can create fire tornadoes inside the fire itself. The air inside the firestorm becomes superheated, just like a forced air furnace – it can melt steel, turn roads into flammable rivers of tar, and spontaneously ignite anything in it's path. The superheated air is then thrown into the atmosphere, generating large clouds, called pyrocumulonimbus clouds.

Eyewitness accounts from the town of Peshtigo itself tell of children flashing into flame, even as they ran toward the river that bordered the town. The river proved some help, but the frigid waters drowned some and caused hypothermia in others. The only people left alive at the end of the storm were people who could stay submerged under the water, then surface for a quick gasp of air filled with soot and flame, and then go under again.

Firestorms are not very well understood, but scientist and meteorologists have called them "nature's nuclear bomb". We know they require a series of several small fires to converge in a specific way. We know that they require a minimum area of about a half of a square mile to develop. We know that temperatures inside these firestorms can reach up to 2000 degrees Fahrenheit. We know that they are self-limiting for a very unusual reason.

The unique structure of a firestorm means that once the critical mass has been achieved, the fire has difficulty spreading. As the winds coming in from outside the firestorm feed it, they also keep the fire from spreading farther. There are exceptions – winds can cause surrounding vegetation to become desiccated, while the radiant heat can spontaneously ignite it. Additionally, tornadoes created inside the firestorm (called firenadoes) can spin away from the firestorm itself and ignite surrounding areas. Although the Peshtigo fire has been largely forgotten by the general public, it's been well studied and documented by two different groups of people, for two very different reasons.

The first is the American military, who studied the effects and origination of the Peshtigo firestorm in order to recreate it during the bombing of Hamburg, Dresden, and other German and Japanese military targets with incendiary bombs during WWII. Creating a firestorm meant that only people and structures in the immediate vicinity of the planned firestorm would be affected – the fire couldn't spread to engulf the rest of the city, or in some cases, the Allied troops on the outskirts of the cities. These campaigns were exceptionally successful, although arguably not so much for the civilians who happened to live inside the bombing areas.

The other group of people that have studied the Peshtigo fire are the forestry and firefighter professionals that work the California wildfires today. Unlike the military, who were more concerned about starting a firestorm and less concerned about stopping it, the people that study the California wildfires are more interested in preventing them.

Although it's been underreported, California wildfires are growing in intensity and have been for the past few years. In 2017, wildfires killed 45 civilians, and 2 firefighters. Wildfires burned over 2100 square miles of California, 250 square miles more than the Peshtigo fire of 1875. That's an area over 4 times the size of Los Angeles.

This, in itself, is horrifying, until you consider that the 2018 season was significantly worse. The 2018 wildfire season saw nearly 3000 square miles of California burned, with a loss of 97 civilians and 6 firefighters. Of course, it's not just California – in Europe, Greece had the Attica wildfires in 2018-2019 that killed 108 people.

However, the wildfires that overtook Australia in late 2019 and into early 2020 have been the worse wildfires the world has ever seen. Nearly 20,000 square miles of land has burned to the ground, and an estimated one billion animals have been killed as of this writing. On Jan 6, 2020, the Australian Prime Minister said that "The fires are still burning. And they'll be burning for months to come."

This is the new normal in California and the rest of the world, and it's getting worse. Wildfires in California are driven by the Santa Ana and the Diablo winds. The dry, mountainous regions are already tinderboxes, but climate change makes vegetation drier with increase of temperature and the subsequent decrease in moisture. These conditions are generating feedback loops that make these areas even more susceptible to wildfires. Home insurance companies are starting to classify homes in certain areas of California as 'high-risk" and denying coverage to homeowners.

The Camp wildfire of 2018 is probably the most well-known fire in recent years, mainly because of the press coverage of the complete evacuation and subsequent destruction of Paradise, California. The debris from the fire required a significantly larger cleanup than the one generated from the World Trade Center after the 9/11 attacks. The fire saw 9,800 homes burned to the ground, along with businesses, hospitals, and schools. Evacuations saw thousands of people sitting in traffic jams for hours while the fire raged behind them. A year after the fire, only 11 homes had been rebuilt.

At the same time the Camp fire was burning, the Woolsey fire saw press because of the people who were forced to evacuate. Malibu, home of some of the entertainment industry's most highly paid actors and musicians, was threatened by this wildfire, and many lost their estates to the flames. In this fire, nearly at the doorway of Los Angeles, embers were blown across the 8 lane Pacific Coast Highway to ignite the other side and continue it's advance.

The Camp fire has been proven to have been started by sparks from electrical transmission lines. As a result, in 2019, Pacific Gas & Electric began rolling blackouts, ostensibly to limit the possibility of starting wildfires from downed lines or sparks from transformers. However, it's been theorized the PG&E has let it's electrical infrastructure crumble over the past several years in order to maintain profits for its shareholders.

So now, California residents are stuck between a rock and a hard place. They can suffer rolling blackouts during the fire season (which runs from April to November), or, they can risk fires starting from electrical transmission lines that have been downed in high winds. Incidentally, the Camp wildfire has not been the only one in California to have been started this way.

In Scenario #1, we talked about the destruction of Miami from a rising sea levels and a hurricane hitting the exact right place. Here's the first vector of Scenario #3 – a series of smaller California wildfires that erupt into a firestorm. Whether the fire causes fatalities or simply causes hundreds of thousands of evacuations and permanent migrations is immaterial. The point is that large areas of California are going to be uninhabitable because of the continued risk of these fires, the inability to get homeowners insurance to cover replacement costs of home building, and the total devastation to infrastructure left behind by the incredibly high temperatures of the firestorms.

California has the most people of any state in the nation, and it's the third largest in terms of land size. One person in four living in California is at high risk to wildfire. That's approximately 10 million people. They don't know it yet, but these Californians will join the ranks of climate refugees all over the nation, and the world.

Our second vector in this scenario starts with a problem that has been growing in the southern United States since the 1980's. I'm talking about feral hogs, and they are a major nuisance in all the southern United States. Most of California, Florida, and Louisiana have feral hog problems, but Texas has by far the largest estimated population of feral swine, with somewhere around 1.5 million feral hogs in Texas.

"Feral" animals are classified differently than "wild" animals. A feral animal is one that was previously domesticated, or it's recent ancestors were domesticated. These animals either escaped, were turned loose, or were abandoned, and became non-domesticated. Most people don't know it, but swine are not indigenous to North America. The first person to bring pigs to the New World also brought smallpox – Columbus.

A feral sow reaches sexual maturity at 6 to 8 months of age, and can have litters up to 14 piglets. She can have a litter 1.5 times per year. Pigs are regularly classified as one of the most intelligent animals on the planet – they are as smart as most primates, and significantly smarter than dogs. Most animal experts agree that they exhibit problem solving skills equal to a three year old human. They are social creatures that care for and protect their young. Pigs remember where food is, how traps work, and hunting grounds – in other words, they adapt to human hunters and learn from their patterns.

Feral hogs have become an expensive nuisance for farmers, land developers, and other people in rural and suburban areas. They are opportunistic omnivores, which means they will eat almost anything, including deer corn, gardens, and even other animals (wild hogs are not affected by snake venom and have been known to harass and eat rattlesnakes). Corn farmers have seen their seedlings methodically eaten directly after planting, and hay and cattle farmers have seen whole fields rooted up for acorns and pecans by their powerful snouts. Wild boar routinely tear up golf courses, sports fields, and will trample through creeks, streams and other waterways. They destroy hay fields for grazing cattle and other native plant life.

The problem has gotten so bad in Texas that in 2019, Governor Greg Abbott signed a bill into law allowing "any landowner, landowner's agent or lessee" the ability to hunt wild boar without a license in the State. In other words – anyone.

Experts agree that the feral hog problem in the South can't be eradicated, at best, they can only hope to contain it. Poisoned bait won't work for feral hogs – they are too smart to eat it and you'd just end up killing the rest of the legitimate wildlife. Even with no restrictions and no licensing needed to hunt in Texas, and less and less restrictions in other Southern states, the feral pig population seems to be doing just fine.

On the other side of the spectrum, consumption of domestic swine in the United States is at an all time high, and pork has long been the most popular meat in the world. The average American eats 193 pounds of chicken, beef, and pork every year. Out of that, 52 pounds are pork, including bacon. That means that fully a quarter of all the protein the average American takes in annually is from pork. Every year in the United States, 120 million pigs are slaughtered for meat.

Pigs are used for a variety of different cuts of meat as well as other products. Your Christmas ham and morning bacon are made from domestic swine farms. Bacon, of course, comes from pigs belly. Cooking lard is made from the fat of domesticated pigs. Suede for shoes and clothing is made from pig leather. Most people don't realize that insulin is made from pigs, as well as valves for human heart surgery. Gelatin is made from pigs. Swine by-products are used in antifreeze, floor waxes, crayons, chalk, adhesives, as well as soaps and make-up.

Domestic swine farmers are probably at the most risk of financial damages from feral hogs, because feral hogs, by their nature, are vectors for disease. According the USDA, feral hogs can transmit a variety of diseases to domestic hogs. Usually contamination occurs near fences, but feral males can and do mate with domestic females through fence breaks. Area where feed or water are stored, as well as creeks or rivers that go from an unfenced property to a fenced one can also be transmission vectors.

Right now, the diseases that feral hogs transmit to their domesticated brothers and sisters are: Brucellosis, a bacterial infection that affects breeding; Pseudorabies, a virus that affects breeding and general health: and Porcine Reproductive and Respiratory Syndrome, a virus that also affects breeding and general health. None of these are fatal and are relatively easy to control.

However, there is a problem brewing in other parts of the world. A hemorrhagic fever called African Swine Fever has been spreading in North and South Korea, China, and from the European Union into Russia. This virus, while harmless to humans, currently affects more than 55 countries on three continents. Infected pigs have near 100% mortality within 7 days of infection. It's transmitted in feces, by ticks, and eating infected pork products.

African Swine Fever first appeared in Kenya in 1907, and stayed restricted to the African continent until an outbreak in Portugal in 1957. From Portugal, the disease spread to Spain and France. The disease was finally eradicated in the 1990's in these areas through widespread culling.

In 1971, it made its way into Cuba, where 500,000 swine had to be killed to prevent a nationwide epidemic. It's been reported in the Dominican Republic, and other areas in the Caribbean Sea. Asia has seen very recent outbreaks, including both South and North Korea, China and Vietnam. In February of 2019, Vietnam reported their first case of African Swine Fever. By October of that same year, Vietnam had culled more than 5.7 million hogs. Their normal production of hogs annually is about 28 million. This means that Vietnam had to cull 20% of its swine population to attempt to stop the spread of the disease, and they still have not been successful at stopping it.

As we mentioned before, America's swine population is about 120 million hogs. The US has not had an outbreak of African Swine Fever - yet. If and when it does, the US swine market will be directly in the crosshairs of what could become a very expensive and deadly problem in the United States food supply.

Due to the feral hog problem the United States, an epidemic of African Swine Fever would spread from the beaches of the Atlantic to the Pacific in a remarkably short period of time. The worst case scenario is the infection begins in Texas, but an outbreak anywhere inside our borders would probably cover the southern half of the United States in less than 3 years.

The feral hog population would be the transmission vector that would spread the virus over a large geographic area. However, because of proximity and cross contamination to domestic hog farms, infected feral hogs would easily transmit the virus to domestic hogs. The entire United States swine industry could see a major contamination of this hemorrhagic virus.

Imagine the culling of 20% of the 120 million hogs in the United States. That's the destruction of 24 million pigs. Assuming a hog is worth about $600 at slaughter, that's a loss of $14 TRILLION dollars. This is only the lost market value – it doesn't include the cost of destroying, burying, and decontamination of this many pigs and the farms that they came from.

This type of zoonotic pandemic would affect every single American. Remember than Americans eat an average of 53 pounds of pork per year. This type of pandemic would reduce the available protein to Americans by 10 pounds per person annually. This doesn't mean that the average American would just shave 10 pounds from their meat diet. It means they would move to another type of protein – either fish, chicken, beef, or plant-based proteins.

The need to fill this gap on the meal tables of American families will cause the cost of other protein sources to rise. The entire food industry would also have to bear the financial burden of the culling and biological cleanup of an epidemic of this size. Inevitably, food costs would rise. Even a 10% rise in food costs would put nutritional meals out of the budget of a significant portion of American families.   A family of four in the US spends an average of $12,000 a year on groceries. Could this same family afford an additional $1200 a year to get the same level of food nutrition? The answer is – not all of them. And of course, it's the poorest people who would suffer the most.

This is the second vector of our last scenario – a zoonotic pandemic that has a major effect on the US food population. The American government is well aware of the devastating effects of this type of pandemic, but not for the reasons you might think. African Swine Fever is one of the diseases that was considered for the American biological weapons program. This program, which was ended in 1973 (at least publicly) researched the possibility of weaponizing African Swine Fever for use as an anti-animal biological agent. This means that the US government knows exactly what mortality and transmission vectors of this virus can be.

While African Swine Fever was only a consideration for a biological weapon, the military did succeed in weaponizing human diseases, other anti-animal diseases, and several anti-agricultural agents, such as wheat stem rust spores and rice blast. Fun fact: there have only been two viruses effectively eradicated from the world as we know it. One of them, of course, is smallpox. The other is a cattle disease called Rinderpest – which is a hemorrhagic fever similar to African Swine Fever, but that affects cattle. Rinderpest was successfully weaponized as part of the American biological weapons program.

Although neither the looming prospect of California fires or a severe hit on American pig market would push us into the tipping point to the next human die off, the combination of the two would definitely stretch the resources of our government and emergency management departments, leaving an opening for the third vector. This last vector is the most frightening of all, because we, as a species, are getting closer and closer to literally unfreezing it.

The third vector to this scenario is frozen somewhere in the world's permafrost. Permafrost, by definition, is ground that stays frozen for two years or longer. It is found in northern Russia, Canada, Alaska, and Finland, as well as other countries above the arctic circle. It's also found at very high altitudes and at extreme southern latitudes. Permafrost can go hundreds of feet deep into the earth, stopped only at the line where geothermal energy reaches above the freezing point.

Permafrost thaws and freezes according to cycles that are dictated by a variety of factors. Seasonal thaws can occur on the top level of a geographical area, but the earth just a few inches down could stay frozen and still meet the two year minimum.

Of course, we already know that cycles of thaw and freeze can be on a much longer scale. The last ice age ended about 10,000 years ago, and the line of demarcation for permafrost has been moving steadily towards the poles since then. Of course, with climate change, thaws are occurring more and more often, and at higher (and lower) latitudes.

As permafrost rises above the freezing point, the microbes in the soil unfreeze as well. Some of these bacteria produce CO_2, and some anerobic bacteria produce methane. We already know that these greenhouse gases accelerate climate change. This positive feedback loop is already in place and happening as you are reading this page.

We also know that the dead and frozen vegetation in permafrost holds a massive amount of CO_2. As this organic matter thaws, it continues to rot and enters the atmosphere as CO_2. Sometimes it's harvested as peat, and burned as fuel, adding more CO_2 to the problem.

Scientists know that permafrost holds many secrets to our past not only as a human species, but also to other animals and plants. In 2012, Russian scientist were able to germinate seeds found in a prehistoric squirrel burrow. These scientists were able to germinate and grow a plant that has been extinct for over 32,000 years.

This phenomenon goes well beyond microbes and plant life. In 2019, the remains of a puppy frozen in permafrost 18,500 years earlier were found. This canine cannot be definitely determined as a wolf or a dog by scientists, leaving some to consider that the break between wolves and domesticated dogs may have happened around this time period. In 2015, two cave lion kittens were found that were over 10,000 years old, proving for the first time that these animals actually existed outside of cave drawings. In 2003, in the Canadian Yukon, a horse was found that was over 700,000 years old. The DNA sequenced from this sample is the oldest genome ever sequenced on earth.

However, there is a potentially devastating problem lurking in northern permafrost. This problem deals with more than just climate change emissions or random specimens being discovered. it concerns human, animal, and plant pathogens that have been stored in permafrost for thousands of years, and perhaps tens, or even hundreds of thousands of years.

In 2016, in a remote area of Siberia, an outbreak of anthrax infected nearly two thousand reindeer. This outbreak killed one young boy, caused dozens of people to be hospitalized, and caused the Russian government to airlift some of the citizens from infected areas. It's thought that the outbreak was caused by the corpse of a reindeer that was killed by the disease and encased in permafrost nearly a century ago. A recent heat wave due, in part, to climate change thawed the carcass, and contaminated the surrounding area with anthrax spores.

There's more to consider as well. We know that the Alaskan and Canadian permafrost holds graves for victims of the 1918 Spanish Flu epidemic. In Siberia victims of both smallpox and plague have been buried for centuries. As permafrost melts, these graves also melt. What's worse is that in permafrost, the ground actually moves as it melts, regurgitating shallow graves due to methane gas expansions as well as the freezing point of water with and without salts, sediment or other organic materials.

Scientists know that above 5.3 million years ago, there were dense spruce and pine forest above the arctic circle. About 2.6 million years ago, geological changes forced a thermal isolation of artic waterways and began the series of feedback mechanisms that would lead to runaway cooling – the opposite of what we are experiencing now. This cooling caused glaciers as far south as present day St Louis, New York City, and Seattle.

Here's the point, and it's far more science than it is science fiction. There are parts of our planet that have been frozen for the last 2.6 million years, and we are in the process of un-freezing them. Humans have only existed on the planet for about the last 100,000 years, and only about the last five thousand is documented history.

In other words, we are in the process of thawing ground that has been frozen for 96% longer than humans have existed. What if a pathogen is released through melting permafrost that we have no way of treating, vaccinating, and have no natural immunities from?

It wouldn't even have to be a human pathogen, in fact, chances are that it wouldn't be. However, a pathogen that affects a specific type of animal – for example, fowl – could be devastating to the human food chain. Likewise, a pathogen that affected crops could be a global problem. Imagine, for example, a pathogen that infected soybean, corn, or wheat plants. These are the some of the most imported and exported crops in the world.

There are other instances of invasive pathogens wiping out animal populations. Right now, a fungus called "White Nose Syndrome" is wiping out bat populations all over North America. Bats are extremely important in keeping the insect population in check. Worldwide, a fungus called Chytridiomycosis has spread from its home on the Korean Peninsula to the rest of the world, infecting over 500 frog and amphibian species. Out of these species, nearly 20% are now extinct in the wild due to this disease.

How quickly could a pathogen spread? Let's look back at African Swine Fever. From our friends at Wikipedia:

- Since around 2007 to August 31, 2018, 1367 cases of African Swine Fever of domestic pigs or wild pigs were reported by veterinary department of the Rosselkhoznadzor (Russian: Россельхознадзор), a Russian federal agency that supervises over agriculture, and state media. According to official report the central and south districts were among most affected by the disease (with several occasions on the east).
- In August 2012, an outbreak of African Swine Fever was reported in Ukraine.
- In June 2013, an outbreak was reported in Belarus.
- In January 2014, authorities announced the presence of ASF in Lithuania and Poland, in June 2014 in Latvia, and in July 2015 in Estonia.
- Estonia in July 2015 recorded its first case of African Swine Fever in farmed pigs in Valgamaa on the country's border with Latvia. Another case was reported same day in Viljandi county, which also borders Latvia. All the pigs were culled and their carcasses incinerated. Less than a month later, almost 15,000 farmed pigs had been culled and the country was "struggling to get rid of

hundreds of tons of carcasses". The death toll was "expected to rise".

- Latvia in January 2017 declared an African Swine Fever emergency in relation to outbreaks in three regions, including a pig farm in Krimulda region, that resulted in a cull of around 5,000 sows and piglets by using gas. In February, another massive pig cull was required, after an industrial-scale farm of the same company in Salaspils region was found infected, leading to a cull of about 10,000 pigs.
- In June 2017, the Czech Republic recorded its historically first case of African Swine Fever.
- The Czech Republic in Zlin, inroduced a regulation that prevented the spread of the ASF infection by removing the contaminated zone via odor fences. Odor fences with a total length of 44.5 km were able to keep the wild boar in the health zone.
- In 2018, Romania experienced a nation-wide African Swine Fever pandemic, which prompted the slaughter of most farm pigs.
- In August 2018, authorities announced the first outbreak of ASF in Bulgaria. By July 2019 five Bulgarian pig farms had had outbreaks of African Swine Fever.
- In July 2019, authorities announced the first outbreak of African Swine Fever in Slovakia.

If you extrapolate this data, it means that a viral pathogen spread by feral, four-legged animals contaminated an area nearly 1200 miles wide east to west, and 1200 miles north to south in just under six years. In case you were wondering, the northern tip of Estonia is about 600 miles from the southernmost permafrost in Russia, and it's all connected by land.

Note that this pathogen was spread by four-legged animals transmitting it without benefit of cars, trucks, or airplanes. Inside the 1200 mile radius just discussed there are hundreds of domestic airports, several international airports, and two major world seaports – Helsinki, Finland, and St Petersburg, Russia. If pigs could do it in six years, how quickly could a two-legged animal with access to all the transportation methods of the 21st century transmit a pathogen? How far could it spread?

African Swine Fever spreads very quickly, and has a near 100% mortality rate. Scientists know some of the vectors of ASF, but not all – it's thought that ticks or other biting insects can also be transmission vectors. The fact of the matter is that any unknown pathogen could have a variety of specific transmission vectors that would have to be researched and eliminated to stop the spread of any disease. Consider that it took nearly 500 years for people to realize plague was spread by fleas.

This is the third and last vector of this scenario, and it's most troubling because a pathogen released from permafrost could be anything. If it's a human pathogen with a 90-95% mortality rate and is airborne, then we don't really have to worry all that much, as the problem of overpopulation and climate change will come to a sudden and abrupt end. On the other hand, it could be a bacterium that thrives on keratin, which would mean that every single vertebrate on the planet would be in danger of losing their skin, hair, nails, horns, hooves, and feathers. It could be a virus that disrupts the fertilization process of chicken eggs, meaning that the source of 50% of the world's protein food supplies would be at risk. It could be a fungus that disrupts chlorophyll production in plants.

We are opening Pandora's box in permafrost. The thing most darkly funny about the whole problem is that we, as a species, have reached a level of development where we can destroy the checks and balances the planet has put in place to maintain our survival, but we are not yet at the level to fix or reverse this destruction. If that sounds too vague, let me narrow the thought – we have the technology to inadvertently spread a contagion, but not the technology to stop one.

Scientists have proven over and over again that our planet is fragile, and that our biosphere is an interconnected web of cause and effect. An unknown pathogen released from melting permafrost, even if it didn't directly target humans, could lead to a cataphoric chain of events that threatens our food sources, our water supplies, or a variety of other factors that would all have a negative outcome to the animals that inhabit this biosphere – that is: you and me.

These are the three vectors of the last tipping point scenario – the entire western seaboard of the United States on fire, causing the most populated state in the US to be deemed uninhabitable; combined with the unstoppable spread of a known virus in the nation's production of pork, causing a 10% to 25% loss of food protein to Americans. Finally, an unknown pathogen spread from melting permafrost with unknown repercussions, symptoms, pathology, and transmission vectors. Combine these three vectors, and the tipping point to the next human die off has begun.

Other Factors That Could Start the Tipping Point

The factors in the above three sections aren't the only thing that could cause the tipping point to the next human die off. There are plenty of other factors in the United States and worldwide to combine and cause a tipping point. As we've said, the world is a reactionary place, and our biosphere is exceptionally fragile and diverse. Changes we make to our biosphere create a chain of events whose results are often impossible to foresee.

Much like our biosphere, political events create changes whose outcomes are impossible to guess at. We are an interconnected world. Who would have guessed after WW2 that China would become the largest manufacturing country in the history of the world? It's been 30 years since the fall of the Berlin Wall, but communism remains alive and well in China, and they still outproduce us.

These changes, both biological and political, create ripples that have unintended consequences. There are several hotspots of dissention in the world that could be a part of a global tipping point to the next human die off. We've mentioned just a few, but there are more – and if you consider unintended consequences, there are almost of infinite number of factors that could inadvertently push us towards the tipping point.

Venezuela is a major problem, although most Americans know very little about the country, or even where in the world it's located. According to the United Nations, as of June 2019 over four million people had fled the country in the wake of Venezuela's economic collapse. What's darkly funny is that Venezuela sits on one of the largest oil reserves on the planet, larger even than the Middle East. Venezuela is unbelievably rich in gold and other minerals and metals as well. The fact that they are pumping immigrants into neighboring countries, Mexico, and the United States means that the resources there are being squandered by people who are keeping all the money for themselves amid political and police corruption.

Venezuela is a political tinderbox. The fact that America has been meddling with elections in Latin American for decades has not left many national leaders in South America with warm fuzzies for the United States. While Venezuela isn't nuclear capable, they are a thorn in the side of the United States because the government structure is Socialist. Because of this government structure, a blind eye is turned to things like illegal and dangerous mining practices, deforestation, and other environmental factors, as long as the right people are getting rich in the process.

But Venezuela has nothing on Chile and Argentina. These two countries control a significant amount of the world lithium production. While Australia is the largest world producer of lithium, and supply for lithium as well as other rare-earth minerals and metals like cobalt and graphite are keeping up with current demand, the need for the materials used in Lithium-Ion batteries is going to escalate in the coming years.

Because of the enormous value of these limited resources, atrocities will follow the mining and distribution of these materials, especially if these resources are located in third world countries. Right now, nearly 60% of the world's cobalt comes from the Congo, where child mining, human rights violations, and mining safety and environmental concerns have a long and sordid history. With steady increase in popularity of Tesla, and other electric cars and trucks, we are going to see these areas racked by civil wars, humanitarian atrocities, and illegal and destructive forms of mining.

There is another potential catastrophe brewing that has stayed off the proverbial radar, this time in India. India leads the world in antibiotic usage, with their total use more than doubling between 2000 and 2015. This is bringing a world wide focus to the potential calamities of antibiotic resistance.

Bacterium, like all living organisms, evolve to survive from various stimuli. As a species, we've suffered from bacterial infections from diseases like pneumonia, to wound infections, to common toothaches. It's difficult to fathom that up until about 200 years ago, death from dental abscesses were not only common, they were the 5th or 6th leading cause of death for centuries. The discovery of penicillin in 1928 changed all that. Now, antibiotics are prescribed for any sign of a bacterial infection. While there is no question these are life-saving drugs, their overuse has contributed to new, evolved bacterium that are antibiotic resistant.

Methicillin-resistant Staphylococcus aureus (known as MRSA) is a bacterial staph infection that has developed resistance to multiple types of antibiotics. It was first discovered in the United States in 1968. Originally a problem at hospitals and other healthcare facilities, in the 1990's a "community associated MRSA" emerged that predominately affect people that live in close quarters, such as soldiers, athletes, or child care workers.

As antibiotic use, and abuse, increases in countries like India and Mexico, the chances that antibiotic resistant bacteria will develop and spread becomes ever higher. After hundreds of thousands of years of evolution, it took only 40 years for staph bacteria to develop into an antibiotic resistant strain, and only another 30 years to spread to the general public. We are already seeing antibiotic resistant strains of tuberculosis. What would happen if an antibiotic resistant form of pneumonia developed and spread? Or a resistant form of cholera? Or tetanus?

One of the other serious concerns that could cause a tipping point in the near future is our addition for technology. Science is a wonderful thing, and it's allowed *homo sapiens* to flourish in ways that were simply unheard of even fifty years ago. Science will pave the way for our next biological evolution, which is a topic we'll discuss in my next book.

But science, like all things, has it's setbacks, accidents, and mistakes. Marie Curie died from her study of radiation. Penicillin was discovered because of sloppy lab practices. Thalidomide created birth defects in thousands of children, and caused hundreds of thousands of pregnancies to be stillborn. The anthrax deaths from a little-known biological weapons accident at Sverdlovsk, Russia that killed over 100 people in 1979.

Additionally, while scientists are brilliant in their academic field, sometimes questions that would be classified as common sense tend to be lost. One of the best examples of this is the question that NASA engineers asked Sally Ride before her one week spaceflight. (I wont give the answer – you'll have to look it up for yourself.)

These types of accidents and common sense questions in an isolated lab are one thing. But technologies like gene editing, nanotechnology, and AI applications all have serious implications for the continued success or failure of our future as a species. There are hundreds of apocalyptic movies, books, and television shows regarding these technologies and how they can be mis-used, or accidentally released onto an unsuspecting population – from Crichton's "Jurassic Park" to Phillip K. Dick's "I am Legend", to sci-fi movies like "Transcendence", to creature features like "Mimic".

The reality of these technologies can be far more horrifying than anything already imagined. We're aren't playing with a chemistry set in a basement anymore. These technologies are the building blocks of life on this planet, and playing with the concepts of genetics and consciousness is tantamount to playing God. I'm not saying we should stop experimenting, I'm saying that we should carefully consider the implications of our actions as we continue forward. If we don't, I think that it will be a factor of luck, far more than skill, that will keep these technologies from causing a tipping point to the next die off.

The final factor that could cause a tipping point is one that I've avoided discussing throughout most of this book, and that is the collapse of the United States economy. The world relies on the American dollar, and a large enough economic downtown in the US could absolutely cause a tipping point in the United States and the rest of the world.

I'm not an economist, and I don't pretend to be one. In addition, I feel that the complex balance of banks, the federal reserve, the global stock market, the GDP, and the fiscal budget are interconnected in ways that even the most seasoned financial veteran has a difficult time unravelling.

However, I don't have to be an economist to see some obvious signs that American finances are in serious trouble. The deficit of the United States Federal Budget will surpass one TRILLION dollars annually in 2020. In 2016 (the most recent date that Census data is available), over 52% of American children resided in households where at least once person received some form of government assistance.

At the same time, corporations are paying less in corporate taxes than ever before. The Tax Cuts and Jobs Act, signed by Trump in 2017, slashed corporate taxes from 35% to 21%, as well as enacted several loopholes that corporations have used to drop their taxes to zero. In most cases, the new tax laws garnered the largest companies in the United States either tax credits or outright rebates.

There are several examples of this. John Deere, a company that earned $2.15 billion in US income, and reported global profits of $2.37 billion, filed for $268 million in tax rebates in 2018. Chevron made $4.5 billion, and received a rebate of $181 million (and that is before oil and gas subsidies). IBM made $500 million and received a tax rebate of – wait for it - $342 million. That's a tax bill of negative 68% of their profits in the United States.

Guess who is paying for that shortfall? In 2016, 47% of US tax income was from individual tax returns, 33.5% was payroll taxes, and 11% was from corporate taxes. In 2019, only 7% of the tax income came from corporations. The remaining difference of 4% was added onto the people – individual income tax increased to nearly 50% of the US Government tax base, and the remaining 1.5% came from payroll taxes.

These economic policies simply are not sustainable. Neither an individual nor a government can spend their way into prosperity. Sooner or later the bottom will fall out, and when it does, we are looking at a financial crisis that will make the housing and banking crisis of 2008 look like a kid that just lost $20 out of his pocket. When that happens, the wealthy retreat to their bunkers. Where will you be?

A Footnote – Nuclear Detonations

Anyone born after 1945 has lived in the shadow of nuclear war for their entire lifespans. Earlier in this book, we talked about MAD – Mutually Assured Destruction, the principle the United States has used for nuclear policy for the last 50 years. Most of the developed world has a vague understanding of what a nuclear detonation would mean, but there are a lot of things about nuclear war that most people don't know. There are experts that feel that we are closer today to a nuclear war than we ever have been.

I haven't included a nuclear exchange in our "scenarios" because it's not a tipping point, it's an end point. However, I did want to discuss a few things that you might not know about nuclear weapons.

If there is a nuclear detonation on US or Russian soil, even an accidental detonation, that's pretty much the end of the whole ballgame. Both the US and Russia have employed "dead hand" weapons systems, also called "fail-deadly" systems. These systems use a variety of triggers to launch nuclear weapons, including seismic triggers, light, radioactivity, and overpressure triggers to launch a full-scale attack – *especially if military command is unresponsive.*

These systems are designed to launch everything the combined countries have at pre-arranged target packages. There are 3750 active nuclear warheads in the world today (from a high of

70,000 in 1986). The majority are overseen by the US and Russia. This is still enough to kill every living thing with a backbone on the surface of the planet many times over.

Most weapons systems use multiple, smaller weapons for a geographic target. It's harder to get a larger ICBM through a weapons defense system than it is to get several smaller weapons or a weapon that separates into bomblets. As a result, a target won't be hit by one huge weapon, such as the Tsar Bomba (at 100 megatons, the largest nuclear weapon ever designed), it will be hit by several smaller weapons. A declassified target package once showed up to 60 warheads targeting Moscow at one time.

You would think that hitting a city with multiple weapons would be counterproductive, but that's not necessarily true. The blast wave of a weapon can be deflected by geographic features and buildings. If you are behind a large building in a nuclear blast, you have a better chance of surviving because of the deflection. After the primary detonation, the blast wave of a second detonation will have a greater radius due to buildings already being leveled.

Due to the concept of Mutually Assured Destruction, there is not a country in the world that has an active aid plan in case of a nuclear detonation. Not one.

Even if MAD wasn't a factor, it would be incredibly difficult and a waste of human resources to enter a city that's been hit by a nuclear weapon. Sending aid workers, first responders, or doctors into an irradiated city just means that the aid workers are exposed to a lethal amount of radiation. As a result, the military will not send in aid workers into a city that has been hit by a nuclear weapon. The city and it's residents are considered a total loss from the point of detonation to the outer radius of 1 PSI air blast.

What does this mean in a practical sense? To the residents of any American or Russian city, not much. It takes approximately 30 minutes for an ICBM to travel 10,000 miles. Any real exchange of nuclear weapons between the United States and Russia would be over in 2 hours – 4 hours at most. However, if you happen be living in another nation in the world, or be visiting a city that is hit by a nuclear weapon, it means a lot of things. All of them are really, really bad.

If you happen to be outside, enjoying the weather within a mile radius of a nuclear detonation, then your troubles are over. What was your body is now part of a superheated ball of plasma. However, if you are in a fortified building – skyscrapers, schools, government buildings, or commercial buildings – your problems are just beginning. Depending on how close you are to the initial detonation, and how high above the city the detonation occurs, you'll probably survive the initial blast to be trapped in the rubble of the building you are in. If you are underground, either in a subway system or the basement of a building, it's worse – the rubble from the blast will block all entrances and exits to the surface, and the EMP (ElectroMagnetic Pulse) from the detonation will kill all power.

Most people will die of shock from their injuries while trapped, or from dehydration. There will be some survivors who will be able to escape the rubble and begin walking to safety. They will experience "black rain". After a nuclear detonation, the rapid heating and cooling of the air produces condensation. This creates rain that collects radioactive particles and soot as it travels through the fallout and smoke to earth. This radioactive rain coats everything it touches with radioactive particles, ash, and soot. Most survivors exposed will receive a lethal dose of radiation from this rain.

If you are able to escape the rubble, and take shelter from the black rain, you still have to walk through miles of a burning, devastated urban area to make it to safety. Even so, "safety" will be a resettlement or refugee camp outside the city. In a developed country, the military might be able to provide shelter, food, and medical staff. In an undeveloped country, these camps will turn into places to die.

Here's the worst part of a nuclear attack on a major city. In almost every other catastrophe or natural disaster, you can count on someone coming to help you. Earthquakes, landslides, fire, tornadoes – when a city is destroyed or partially destroyed by natural causes, in terrorists attacks, or even in times of war – you can count on first responders. In a nuclear attack, no one comes to help. You are totally and completely alone. In this way, and in many, many other ways, nuclear war is the most dehumanizing of all types of conflict that man has created.

Chapter 7: The First Year – What History Teaches Us to Expect

As humans, our bodies are designed to do two things – survive and procreate. If these are the two demands that evolution has put on our minds and bodies, then a die off due to famine or catastrophic incidents is guaranteed to bring out the worst in human behavior. People driven to desperation will do anything to protect themselves and their families. The first year of the crisis will be the most critical, as food, water, and other resources are going to be extremely limited.

In this chapter, we're going to talk about what history has taught us to expect in similar circumstances. First, we'll look at the effects of starvation on the human body, and we'll also talk about what to expect in areas where mass death from disease is common. We will discuss migration for both the people having to leave their homes, and the areas most likely to see large numbers of refugees.

Effects of Starvation on the Human Body

Any survivalist can quote you the "Law of Threes" – it takes 3 minutes to die without oxygen, 3 days to die without water, and 3 weeks to die without food. If you had to pick one of these, suffocation from drowning or choking is probably the best choice of the three. Dehydration isn't the most pleasant way to die, but it could be a lot worse – once you lose 10% of your body weight in water, your kidneys and liver fail and you've expired. However, death from starvation is... bad.

Death from starvation can occur in famine, but it's also been used as punishment for centuries. The word "immurement" describes a form of punishment where a person is placed in an enclosed space with no exits. Edgar Allen Poe describes an immurement in his short story "The Cask of Amontillado". Roman vestal virgins faced death by immurement if they broke their vows of chastity, and it was used in Persia up until the early 20th century as punishment for thievery. The French even have a term for a type of dungeon used for immurement in the middle ages – they called it an "oubliette" – literally, a "place of forgetting".

Of course, our focus is on famine, and what you might expect when the food resources of a populated area begin to run low. Throughout history, the results of famine are predictable. As food begins to decline, prices go up. Price gouging becomes common. Theft and violent crimes spike in the early part of the famine, but as food become more scarce, and caloric intake drops, people become too weak to try and improve their situation through fair means or foul. In many cases, cannibalism becomes a last resort. Many people will eat inedible items to simply fill their stomachs – shoe leather, grass, and dirt. As death becomes more and more common, disease begins to break out among the survivors from bodies that are unburied, increasing the death toll.

The Minnesota Starvation Experiment

During WWII, the United States, the Nazis, and Japanese were all involved in human testing for a variety of military uses. While many of these experiments were unethical, illegal, and downright horrifying, they did produce data of varying usefulness on the effects of low atmospheric pressure, frostbite, weapons efficiency, biological agents, and antibiotics on humans. (The trading of this information also kept Hirohito from swinging at the end of a rope for war crimes – but that's another topic for another book.)

One of the more useful experiments conducted by the US started in November of 1944 at the University of Minnesota. Thirty-six men volunteered for a yearlong experiment on the psychological and physiological effects of starvation on the human body. This study, called (appropriately enough) the Minnesota Starvation Experiment, was, and to this day is considered the landmark study on human starvation.

The study protocol required that the men lose 25% of their body weight. During the first three months, the men were able to eat a normal diet of 3200 calories per day. For the next six months of the study, the men were required to half their caloric intake, down to 1570 calories per day. The final weeks of the project allowed them unrestricted access to food to rehabilitate their weight lost.

During the six month period of semi-starvation, the changes in the men were dramatic. The test subjects saw significant decreases in strength, stamina, body temperature, heart rate, and sex drive. The physiological effects were significant as well. The men in the study reported irritability, fatigue, depression, and apathy.

"Of course." I can hear the reader saying. "Take a man's dinner away and he's going to be irritable." However, the point of the study was to instruct relief workers on how to deal with the millions of starving civilians and soldiers from WWII. Remember, it was 1945, and no one had any idea of how to rehabilitate the starving population of an entire continent. The leading psychologists of the study published a booklet in the same year, titled "Men and Hunger: A Psychological Manual for Relief Workers." It's publication, as well as the full results in their study published in 1950, showed for the first time real data on caloric intake, the psychology of starvation, and what was required to overcome and rehabilitate starving people. It was used as a guide to rehabilitate an entire continent as the

Allies liberated concentration camps, tended for prisoners of war, and rebuilt the agricultural assets that had been destroyed in the war.

An Eyewitness Account from the Irish Potato Famine

The Minnesota Starvation Experiment was conducted with healthy men, in clinical observation settings. The men involved were able to stay clean, had access to fresh drinking water, and had the advantages of modern medicine to treat illnesses. However, people that are victims of a famine like the one that struck Ireland in 1845 had no such advantages.

The potato isn't native to Ireland, but it became a staple crop there after it was brought over from the New World. At first, the plant seemed like a godsend – it grew well in Ireland's climate and soil, had a high caloric value, and allowed the population of Ireland to grow exponentially. By 1840, fully 1/3 of Ireland's population was dependent on the potato for nourishment. When the potato blight struck in 1845, it lit a tinder that would become a wildfire in an unbelievably short time.

The first year of the blight, only 1/3 of the crops were lost. In the following two years, 3/4 of each harvest was lost. Under British rule, Ireland was stuck between a rock and a hard place. The ruling Whig party of Britain at the time didn't really want to help the Irish – in fact, the British felt as if the famine was divine providence, and that God intended the famine as a judgement against the Irish. The chief proponent of this idea was a man names Sir Charles Trevelyan, who was the British civil servant who was actually tasked with the job of Irish famine relief. In his book *The Irish Crisis*, published in 1848, Trevelyan described the famine as 'a direct stroke of an all-wise and all-merciful Providence', one which laid bare 'the deep and inveterate root of social evil'. This ideology, combined with a mindset of 'famine fatigue' populated among the merchant and ruling class of Britain, caused Ireland to suffer far more losses. This mindset would later become key to another series of famines under British Rule, this time in India, where two separate waves of famines caused the deaths of 20 million people.

In early 1847 the *Illustrated London News* hired an artist and writer named James Mahoney to tour the affected area and report on what he saw. The following is a direct copy of his article, as printed by the News:

> *"I started from Cork, by the mail, for Skibbereen and saw little until we came to Clonakilty, where the coach stopped for breakfast; and here, for the first time, the horrors of the poverty became visible, in the vast number of famished poor, who flocked around the coach to beg alms: amongst them was a woman carrying in her arms the corpse of a fine child, and making the most distressing appeal to the passengers for aid to enable her to purchase a coffin and bury her dear little baby. This horrible spectacle induced me to make some inquiry about her, when I learned from the people of the hotel that each day brings dozens of such applicants into the town.*
>
> *After leaving Clonakilty, each step that we took westward brought fresh evidence of the truth of the reports of the misery, as we either met a funeral or a coffin at every hundred yards, until we approached the country of the Shepperton Lakes. Here, the distress became more striking, from the decrease of numbers at the funerals, none having more than eight or ten attendants, and many only two or three.*
>
> *We next reached Skibbereen... We first proceeded to Bridgetown...and there I saw the dying, the living, and the dead, lying indiscriminately upon the same floor, without anything between them and the cold earth,*

save a few miserable rags upon them. To point to any particular house as a proof of this would be a waste of time, as all were in the same state; and, not a single house out of 500 could boast of being free from death and fever, though several could be pointed out with the dead lying close to the living for the space of three or four, even six days, without any effort being made to remove the bodies to a last resting place.

After leaving this abode of death, we proceeded to High-street, or Old Chapel-lane and there found one house, without door or window, filled with destitute people lying on the bare floor; and one, fine, tall, stout country lad, who had entered some hours previously to find shelter from the piercing cold, lay here dead amongst others likely soon to follow him. The appeals to the feelings and professional skill of my kind attendants here became truly heart-rending; and so distressed Dr. Donovan, that he begged me not to go into the house, and to avoid coming into contact with the people surrounding the doorway...

Next morning...I started for Ballidichob, and learned upon the road that we should come to a hut or cabin in the parish of Aghadoe, on the property of Mr. Long, where four people had lain dead for six days; and, upon arriving at the hut, the abode of Tim Harrington, we found this to be true; for there lay the four bodies, and a fifth was passing to the same bourne. On hearing our voices, the sinking man made an effort to reach the door, and ask for drink or fire; he fell in the doorway; there, in all probability to die; as the living cannot be prevailed to assist in the interments, for fear of taking the fever.

We next got to Skull, where, by the attention of Dr. Traill, vicar of the parish (and whose humanity at the present moment is beyond all praise), we witnessed almost indescribable in-door horrors. In the street, however, we had the best opportunity of judging of the condition of the people; for here, from three to five hundred women, with money in their hands, were seeking to buy food; whilst a few of the Government officers doled out Indian meal to them in their turn. One of the women told me she had been standing there since daybreak, seeking to get food for her family at home.

This food, it appeared, was being doled out in miserable quantities, at 'famine prices,' to the neighbouring poor, from a stock lately arrived in a sloop, with a Government steamship to protect its cargo of 50 tons; whilst the population amounts to 27,000; so that you may calculate what were the feelings of the disappointed mass.

Again, all sympathy between the living and the dead seems completely out of the question... I certainly saw from 150 to 180 funerals of victims to the want of food, the whole number attended by not more than 50 persons; and so hardened are the men regularly employed in the removal of the dead from the workhouse, that I saw one of them, with four coffins in a car, driving to the churchyard, sitting upon one of the said coffins, and smoking with much apparent enjoyment. The people also say that whoever escapes the fever is sure of falling sick on the road (the Public Works), as they are, in many instances, compelled to walk from three to six miles, and sometimes a greater distance, to work, and back again in the evening, without partaking of a morsel of food. Added to this, they are, in a great number of instances, standing in bogs and wet places, which so affects them, that many of the poor fellows have been known to drop down at their work."

The Spanish Flu Pandemic of 1918

The largest death toll from a pandemic in history was the Spanish Flu pandemic of 1918. In the space of less than 2 years, 500 million people were infected, and somewhere between 20 million and 50 million people worldwide died from the disease - 675,000 in the United States alone.

The Spanish Flu pandemic was truly a 'perfect storm' for a variety of factors. The first was the fact that the world was at war. World War One was nearing its conclusion, and hundreds of thousands of troops were mixed in barracks and bases all over the world. Troop movements were a major transmission vector for both the first and second outbreaks of the Spanish Flu, killing about 45,000 American soldiers - just short of the 53,402 killed in action during the actual conflict. Allies and enemies alike contracted the disease and spread it to other military detachments, to civilians they came in contact with, and finally, to their homes.

The second factor was medical knowledge of the time. Microscopy had identified bacteria, but viruses are much smaller than bacteria and were unknown in 1918. Medical science at the time was completely baffled with what they were dealing with. Normal treatments for the common cold and influenza didn't work – in fact, made the infected sicker and more susceptible to dying from the disease. Most doctors of the time realized they were dealing with something that was smaller than a bacterium, but not what it was. Influenza viruses wouldn't be positively identified by microscopy until the 1930's.

But the most tragic factor of the Spanish Flu mutation was the way it affected the body. For centuries, flu only killed the very young and the very old, but the Spanish Flu targeted young men and women in the prime of their lives. In fact, the healthier you were on contracting the disease, the faster you succumbed to it. All over England, there were reports of young adults and children going to school in the morning healthy, and dying literally at their desks before the school day ended.

We now know that the Spanish Flu triggered what doctors call a cytokine storm in the body. Cytokines are signaling chemicals that help mobilize immune cells capable of removing infections from the body. A cytokine storm will typically focus in the lungs, causing excess liquid and inflammation in lung tissue, sometimes hemorrhagic inflammation. The stronger your immune system, the stronger the cytokine reaction. Victims of the Spanish Flu literally drowned in their own bloody pus.

It was not a pleasant death. Most people died gasping for breath, sometimes for hours. One of the most obvious signs that a person was infected was a marked cyanosis of the ears, fingers, and other extremities – victims would turn visibly blue because they could not get enough oxygen in their bloodstream. The mortality rate was somewhere between 10% and 20% worldwide. The key to survival of the infection was immediate bed rest once you started exhibiting symptoms. People that tried to 'work through' their symptoms, mostly soldiers on the march, but also medical personnel, had a much higher mortality rate.

In the United States, the war effort was in full force, and many of the medical personnel had been transferred to Europe. There was a critical shortage of nurses and doctors in most major cities in the United States. To make matters worse, President Wilson signed into law the Sedition Act of 1918. This act, which was later repealed in 1920, made it punishable with a 20 year prison sentence to "utter, print, write or publish any disloyal, profane, scurrilous, or abusive language about the form of government of the United States... or to urge, incite, or advocate any curtailment of production in this country of anything or things... necessary or essential to the prosecution of the war." Government propaganda posters and advertisements urged citizens to report to the Justice Department anyone "who spread pessimistic stories... cries for peace, or {who} belittles our effort to win the war."

With this act as law, public health officials were terrified of being quoted by the press as "pessimists" or "reactionary". As the influenza epidemic started to spread in late spring of 1918, public health officials were caught between a rock and a hard place. Faced with not only the interest of American war morale, but also in the very real concern of being thrown into jail for reporting the truth about the influenza outbreak, they did the worst possible thing – they began to lie.

Philadelphia was one of the hardest hit American cities. The public health director, a man named Wilmer Krusen, refused to listen to doctors urging him to shut down the Liberty Loan parade on September 28, 1918. Doctors were afraid that hundreds of thousands of people, jammed into the parade route, would become a significant vector for infection to unexposed people. The parade went forward as scheduled, and it was an unmitigated success – up to that time, it was the largest crowd in Philadelphia's history. Unfortunately, it also turned out to be the deadliest. The incubation period for the Spanish Flu is two to three days. October of 1918 in Philadelphia saw corpses begin to pile up by the hundreds awaiting burial, as there were no coffins. Cold storage plants were used as temporary morgues. Priests drove horse drawn carriages down city streets, calling for people to bring out their dead. Many were buried in mass graves. More than 12,000 Philadelphians died, nearly all of them in the six weeks following the parade.

All across the country, public health officials lied. U.S. Surgeon General Rupert Blue said, "There is no cause for alarm if precautions are observed." In August 1918, the monthly bulletin by the health commissioner of New York, Royal S Copeland, said that "The public has no reason for alarm since through the protection afforded by our most efficient quarantine station, and the constant vigilance of the city's health authorities, all the protection that sanitary science can give is assured. The very mildness of the disease, as reported in Europe, is, in itself, assurance against anxiety on this side of the water." The Los Angeles public health chief said, "If ordinary precautions are observed there is no cause for alarm."

By November of 1918, it was obvious to most Americans that they were not getting the whole truth, and, under pressure, the press and public health officials began providing real information. Most major cities began closing public buildings, schools, and courthouses to restrict the spread of the disease. Handbills were posted advising people to "not spit on the sidewalk" for fear of spreading the disease, and streetcars in San Francisco urged people to leave the windows down and let the air flow through the car. Even Krusen in Philadelphia took quarantine measures in November, and when the third wave of the Spanish Flu hit the United States in January of 1919, it was quickly eliminated.

However, during those dark months of summer turning to fall in 1918, the "can do" attitude of the United States was nearly snuffed out. In the absence of any real leadership, and with no information coming from their elected officials, people retreated into their homes. Trust evaporated. The community of people helping people evaporated. For a brief time during that terrible season, the America we've come to believe in – neighbor helping neighbor, one nation, indivisible – was lost.

The question that this moment in history should raise is this: What happens when the disaster is longer than a few months?

What happens when a quarantine against a contagious agent isn't effective? What happens when the government lies to keep people from "panicking"?

More to the point – would our government lie to us today to keep the general population from "panicking"? The answer, of course; is yes, and yes, and yes. Between August 28 and September 5, 2019, broadcast TV aired 216 segments about Hurricane Dorian. Out of the three major networks – CBS, NBC, and ABC, only one network mentioned climate change in combination with Dorian. "CBS This Morning" aired a segment on the September 4 episode that mentioned climate change. This was the only segment aired on CBS, and not one of the other networks mentioned climate change, despite it being the worst storm to ever hit he Bahamas (Dorian struck the Bahamas on September 1, 2019). Climate experts also noted that Dorian stalled over the Bahamas for nearly two days – a troubling trend that is becoming more common with tropical storms.

The most heartbreaking stories to come from the Bahamas wasn't the storm, it was the reticence of survivors to seek help. Many illegal Haitian immigrants lived in the Bahamas, and they fill the same roles that undocumented Hispanics do in the United States. When aid helicopters flew sorties over the most affected area, aid workers reported that many people were afraid to signal or come to the aircraft. These illegals were more afraid of being deported back to Haiti than they were of living with no running water, no electricity, and no social services, in a place that literally had bodies rotting around them.

Soon, we are going to face a new threat to our nation, worse than anything we've ever had to deal with before. It will come from two fronts – the loss of coastal lands due to sea level rise, and the future rise of American and Mexican climate refugees.

Climate Refugees and What They Will Face

On May 30, 2019, President Trump threatened Mexico with a 5% tariff on all goods imported from Mexico unless the flow of undocumented immigrants across the United States border stopped. Less than two weeks later, Mexico agreed to send it's national guard to her southern border to stem the tide of central and south American refugees coming from Ecuador, Venezuela, and Guatemala. Suddenly, the human rights issues that plagued the US for the first part of 2019 – which saw the United States separating families, putting children in literal cages, and blocking the press from entering American 'resettlement camps' – were pushed onto another government, to a place that most Americans couldn't care less about.

In Mexico's defense, they didn't really have much choice. Mexico is still considered a developing nation, and faced with escalating tariffs with the United States, they chose to take action against immigrants coursing through their country. In September of 2019, after deploying over 20,000 troops to the southern border and instituting crackdowns inside the country that targeted transportation vectors of migrants – mainly freight trains, buses, and tractor trailers, the Mexican government reported a 56% decrease in undocumented immigrants apprehended at the United States border. Only 63,989 people were caught, down from 144,266 undocumented immigrants in May of 2019.

There are two points to be made here – the first is that they "apprehended" 63,989 people – which means that a portion of them were not apprehended and made it across the border. Most estimates on apprehended immigrants put the number at 10% of people attempting to enter the country – which means, by extrapolation, that about 600,000 people made it across the border into the United States. In one month.

The second point is – what happened to all of the rest of the undocumented immigrants that are coming out of Central and South America? They didn't just decide to go back home. In fact, they are now being held in detention facilities in Mexico, rather than ones in the United States. Now, it's not a United States problem anymore, it's a Mexican problem. Abracadabra!

We talked about Hispanic immigrants in Chapter 5, but without a solution, this problem is only going to get worse. Faced with tariffs on a northern border versus an ever-increasing tide of immigrants on her southern border, Mexico will become more and more desperate to stop the flow of people coming from central and south America. This can only create more detention camps, ethnic hatred, and finally, violence towards people who are trying to escape the unlivable circumstances in the places they are coming from.

In the United States, one of the most documented refugee groups in history were the people of New Orleans after Hurricane Katrina made landfall on August 29, 2005. The devastation of New Orleans was another 'perfect storm' of factors that led to it being one of the worst natural disasters on United States soil.

New Orleans actually sits below sea level, and it's protected by a series of levees that keep floodwaters from entering the city. The levees were rated for a Category 3 hurricane, and Katrina hit Southern Louisiana with Category 5 wind speeds. It was like dropping a bomb on the city.

When the levees failed, over 80% of New Orleans was flooded, in some places up to 20' deep. The final death toll was officially stated at 1833 people, but the true number will never be known – even today, there are still people missing from the storm, and there was no way to count the city's homeless population before or after the storm.

Not so incidentally, the United States government reaction to the disaster was unbelievably lax. (Except for the Coast Guard, who were running rescue missions into flood ravaged areas almost as soon as the rains stopped.) Two days after the storm, the New Orleans police force advised their officers to use deadly force against looters (this has been hotly debated and denied by NOPD decision makers). Displaced people congregated in the Superdome and the New Orleans Convention Center, where dwindling supplies caused fights, rapes, and hoarding. Four days after the storm, Bush signed a $51 billion relief bill – but some members of Congress felt that the four days was too long, and that the timeline to grant aid was hindered by the race and economic stability of those affected by the storm. Finally, five days after the storm, FEMA delivered fresh water to the Superdome. Read that again – it took FIVE DAYS for FEMA to get clean drinking water to New Orleans after Katrina.

Because it was a census year, we have the most accurate number of people living in New Orleans possible. There were 484,874 people living in New Orleans before Katrina, and an estimated 230,172 people living in New Orleans five years later, in July of 2006. New Orleans lost half its population due to Katrina, and, as with any crisis that involves a mass migration, there were some people that had the means to move and resettle, and some who were evacuated with nothing but the clothes on their backs.

Although many cities in the south took on refugees numbering in the tens of thousands, Houston became the landing place for an estimated 150,000 people fleeing the devastation in New Orleans. Dubbed "New Orleans West", Houston officials opened up the convention center to refugees, as well as the Reliant Astrodome and other government buildings. This doesn't count other groups that landed in Houston – for example, it's estimated that 9,000 ethnic Vietnamese people evacuated from New Orleans were integrated into Houston by living temporarily in homes owned by other Vietnamese that were already Houston residents.

Interestingly, the reaction to New Orleans refugees by Houston residents was overwhelmingly negative, and most importantly, crossed racial lines. In 2008, the percentage of Houston residents that called the Katrina experience a "bad thing for Houston" hit 70%. This is in a city whose demographic at the time was roughly 50% white, 25% black, and 25% other minorities. Most of the refugees that ended up in Houston were poor and black. This means that a significant portion of the black community of Houston residents went against racial lines.

There were other issues as well. The refugee children that entered the Texas school systems were woefully behind their peers due to the quality of public education in pre-Katrina New Orleans. Government programs like WIC and Medicare had a difficult time handling the sudden influx, and of course, the job market was temporarily thrown into disarray. There was a perception of higher crime among areas where there were more refugees, but this was later proven by police statistics to be nothing more than a rumor.

These problems have resolved themselves over time, but it was far from an easy road. Even today, New Orleans is still at 80% of population levels pre-Katrina. Most importantly, the lessons that the United States has learned under Katrina has been bitter, to say the least. Benjamin Franklin may have said it best: "Guests, like fish, begin to stink after three days". Although humorous in nature, this old saying has serious undertones.

As residents of a developed nation, we've watched the world turn away climate refugees in Central America, war refugees from Syria, and watched the centuries old fight over the Gaza strip destroy hundreds of thousands of lives. But that's not even the real truth, is it? We never even watched these events. We couldn't be bothered. We've ignored these events, even aggressively ignored them. It's been nearly 20 years since the resettlement of Katrina victims, and we haven't learned anything.

Rising sea levels will cause mass evacuations in the United States again – it's really just a matter of time, now. Miami-Dade County has 2.7 million people as of this writing, and is just 6' above sea level. That doesn't include the Keys, or north into Fort Lauderdale. Where are all of these people going to go when the seas rise, and there is simply no home to go back to?

Chapter 8: Learning From History – Practical Applications

Today's modern interpretation of the term "zombie" is based
largely on George Romero's iconic film "Night of the Living
Dead", released in 1968. Of course, cultures throughout history
have all dealt with the concept of a mindless, reanimated
automaton. Some, like the Jewish legend of the Golem, meant
a creature made from clay that was brought to life to do its
master's bidding. The actual term "zombie" comes from Brazil,
but it was the Haitians and their ties to West African voodoo
that truly cemented the concept of the dead rising to prey on
the living in pop culture. It's even reported that the CDC has a
"zombie action plan" in case of a hypothetical "zombie
apocalypse".

In films and television shows of modern America, the typical
zombie apocalypse will have hordes of undead running after
living people. Granted, a corpse chasing after you, even if it's
just moving at a brisk walk, is still a cause for alarm. The signs
of the coming die off won't be that obvious.

In fact, the media channels we've come to rely on for our news
will almost certainly work against us. We've already talked
about media saturation. We've talked about how the channels
we select for our news makes us more insulated against what is
really happening in the world. And frankly, it's against the
better interests of big business for us as a national population to
wake up and think about what's really happening. After all, it's
easier to sell a product to someone who doesn't know there are

other options available. The fact of the matter is that we, as a population, may have become the zombies ourselves.

As I've written this book, I've done my best to keep my observations rooted firmly in historical observations. This chapter is the culmination of everything I've written in the previous pages - the distillation of my research and knowledge on this subject, if you will. *Here is where we make the leap from observation to action.*

Predictions for the Future

There are two questions that you need to be asking yourself at this point of the book. The first is "When?", and the second is "How Long?"

I really wish I knew. The world is a reactive place. Some events trigger changes that happen explosively. Who could have guessed that the assassination of a relatively unknown Austrian Archduke would start a chain of events leading to the outbreak of WWI?

On the other hand, some events, once placed in motion, take decades to unfold. Between 1776 and 1787, the founding fathers of the United States effectively pushed the question of slavery to future generations. It took nearly 100 years for civil war to break out in the United States over the issue of slavery, and the issue wasn't fully resolved until the ratification of the 13th Amendment in 1865. Some would say the issue still isn't fully resolved today.

We can, however, look at historical information and make educated guesses on timelines. The last large pandemic was the Spanish Flu of 1918, which killed 100 million people. The last World War killed 70 million and ended in 1945.

If the assumption is made that climate change is the reactive force behind the next human die off, then scientists have already given us a timeline. Once we go over 3 degrees of warming, we've passed a crucial point of our de-evolution.

So let's take the first of these two concepts: the period of ebb and flow of historical mass deaths, both from a perspective of manmade death from war, and natural deaths from pandemics and famine. Add the future perspective of climate change from scientists who have studied what our carbon use is doing to our planet. If we take these two perspectives and dovetail them together, then we can safely say we have somewhere between 15 and 30 years to be ready for the next mass human die off. Of course, the events that lead us to the tipping point may come tomorrow. If they do, hopefully the events I've predicted in this book will be of the type that take many decades to unfold. However – I doubt it.

I believe that once the events leading to our next mass death begin, unalterable changes will come very quickly. The world moves more rapidly now than it ever has. Information is processed and disseminated faster than ever before. Transportation allows people, products, and services to move faster than ever before. Why wouldn't it make sense that our next die off, once began, will move at the same speed as the rest of the world already does?

Conversely, and even ironically, government moves slower now than ever before. It takes years for government policies to change. Decades for laws to change. It seems that since the end of WWII, federal government service and government jobs are perceived as the least desirable jobs in not only the United States, but also in Europe. Furthermore, the general perception of federal government workers seems to be dullards who couldn't find a better job somewhere else. (I'm excepting local government and the armed forces from this statement – you guys rock.) If our governments no longer have the best and brightest of our population, then where does that leave us in response time to national emergencies? Remember, it took five days for FEMA to get drinking water to the Superdome after Katrina. What will the response time of our governments be at the events leading to coming human die off? We know the government will move fast enough to save themselves. Will they move fast enough to save any of us?

Preppers call this the "SHTF" moment. It stands for "Shit Hits the Fan" and there are people all over the developed world, from all kinds of backgrounds that are actively preparing and planning for it. Some, like the wealthy and ruling classes we've discussed, will run to their fortified bunkers to wait out the conflict. Some will shelter in place. Some have second homes that have been specifically outfitted to last for a period of time completely off the grid. Again, this isn't a prepper book and I'm not suggesting you start digging a shelter in your backyard. What I am suggesting is for you to consider that we have a finite amount of time before the die off begins, and that you might think about some suggestions that will help you survive the coming die off.

The Importance of Rural vs Urban Living

I recently visited southern Ireland with some friends. The small town we stayed in was named Clonakilty, and has been there in its current form since at least the 1200's. Southern Ireland is amazing – it's stunningly beautiful, temperate, and the people are exceptionally friendly. Most importantly, there is an agricultural base that is simply non-existent in the United States.

We've already talked about the Irish Potato Famine in this book, and it's clear that southern Ireland, at least, has kept the lessons from the Famine. It seems like everyone maintains a small garden. Small produce farms, poultry farms, swineries, and cattle farmers are simply everywhere. Grocery stores proudly carry locally baked bread, locally slaughtered beef and pork, and local eggs, milk, and cheese. We visited the fishing village of Baltimore where people have fished their waters for thousands of years.

Of course, even Ireland has had its mistakes. Baltimore overfished its waters in the early part of the 19th century. Currently, the Irish fishing industry is listed as one of the world's worst overfishing suspects. The Famine itself could be considered a mistake – the potato isn't indigenous to Ireland, and their dependence on it created an environment that set the proverbial kindling for the Famine. But Ireland is light years ahead of the United States when it comes to surviving the die off. The ability of the population of southern Ireland to maintain a local agricultural base will be a huge factor in their survival.

In the United States, the opposite is true. We have become a population of consumers. The majority of us don't know how to grow our own vegetables. We have lost the knowledge of our grandparents and great-grandparents when it comes to farming for produce, poultry, or wild meat. For us, beef and chicken arrives in packages at the grocery store. The vast majority of the population of the United States has never had rabbit, duck, or goose – but these were daily meals for people living before the 1900's. Dove, pigeon, and squirrel were all fair game for our ancestors. Turkey is still consumed regularly in the US, but only on a specific holiday, which is a holdover from our colonial forefathers (and foremothers). Thanksgiving was made a national holiday by Lincoln – in 1863, two years before his assassination.

We are approaching a nexus in food availability in the United States, and worldwide. As Malthus predicted, our population growth has been a direct result of food abundance. Now, however, we are starting to see signs of food scarcity. It's estimated that 80% of marine species are overfished. Brazil is burning down the Amazon forest to provide more grazing land for Brazilian beef. Climate change is threatening cropland worldwide and changing weather patterns that provide snowpack and rainfall for lakes and rivers. Fracking is poisoning our groundwater resources.

As food and freshwater resources become more scarce, the laws of supply and demand will cause food to be priced out of the range of normal working people. This also happened in the Great Depression, but there were some key differences in that crisis that are going to be very different from the one we are facing.

In the 1930's, no one starved to death. People went hungry, sure, but no deaths were reported from starvation during the Great Depression. However, in the 1930's, the population of the United States was 123.2 million people. In addition, the rural population to urban population numbers were 43.9% versus 56.1%, respectively. In other words, nearly half the population of the country in the 1930's knew how to grow their own food.

In 2018, the population of the United States is 327.2 million – almost triple the amount of people that lived in the US during the Great Depression. In the United States today, only 19.3% of the population live in rural areas. Only 2% of Americans live on active farms. The other 80% live in urban areas. This means that, at best, only 20% of Americans know how to grow food. At worst, only 2% know how to grow food. This leaves the other 80%-98% to rely on those people to provide for them.

This paints a very black picture for the urban middle classes of today. As food becomes more scarce, and water resources more limited, bringing food into cities from rural areas will become more costly. In our cities, people in middle classes or working classes that have jobs with fixed incomes will start to see their paychecks able to buy less and less. We are already seeing the beginnings of civil unrest due to income inequality. We know that the gap between the poor and the wealthy is widening.

For the first time since the industrial revolution, we are starting to see a population shift to rural areas. These people, largely middle to upper middle class, are the forefront of a much larger shift that is coming. In the coming human die off, people that are living in rural or suburban areas with the ability to grow their own food, or have the availability to trade for food, are going to be much better off than their cousins still left in urban areas.

Skills to Train For and Have When the Die off Comes

Does this mean that high-tech jobs and positions will no longer be available? Of course not. Technology will continue to grow and expand. Social media companies will continue to grow, as will tech companies. Security companies will grow. The internet will continue to grow and expand. These companies, as well as the United States armed forces, will continue to need talented people as technicians. The die off won't change that, and in fact, I feel that technology, and technology companies, will continue to mature as we live through and beyond the coming crisis. We'll talk more about future technology in our second book in this series, but for now, I don't believe we are looking at a world that is dying. I believe we are looking at an evolutionary die off event where the people that survive will live in a truly golden age.

The simple answer is that these companies, like the military, the wealthy, and the ruling classes, will become more fortified during the coming crisis. Our colleges will become more fortified. Food centers will become more fortified. The coming disaster won't stem the tide of technological innovation – it will just be done behind higher and more secure walls. Likewise, people that have these skillsets, or other useful skillsets, will be able to survive and even thrive in the coming disaster we are facing.

One of the more bitter ironies in the coming die off is that people who have embraced our technologies without understanding the hardware or software behind it will be the first to struggle. The urban social media influences, the SEO marketers, the people who seems to live on their phones, texting, tweeting, and endlessly posting their opinions to people who are also endlessly posting about their opinions – these are going to be the first to struggle. Building a computer is a marketable skill. Using a computer to post stories about your cat is not.

It's simple, really. In a race, the fastest win. The coming die off will be a race to secure yourself and your family with the resources needed to survive, and most importantly, to stay ahead of the people who didn't make their own arrangements. One of the best ways to win a race is to prepare and train. In this case, training will include acquiring the skills necessary to survive.

First of all, teach yourself a skill that can be used to trade for food. Let's take for example – an attorney that deals in corporate collections. Once the food starts running out, that person is going to be in a world of hurt. If all you are qualified to do is push paper, then you really should look at diversifying your skill set. As the divide between the wealthy and the poor grows wider, more middle class citizens are going to fall below the poverty level. With scarcity of food and water causing rising food costs, we WILL start to see deaths from malnutrition in urban areas – first a trickle, then a flood.

On the other hand, members of the working class, especially those living in semi-rural areas, will do OK. Consider an electrician, or a good, working mechanic. His neighbors will be able to grow enough food for trade. Someone in the area will have a water well – and if they've planned well, it will be a water well running off of solar cells. People who can trade their skills will be in high demand, and will stay in a middle or working class level.

Am I suggesting that you drop out of law school and start growing your own sweet potatoes? Of course not. But there's no reason you can't do both. Plant an indoor garden – these skills may become invaluable later on, if not for you, then for your children. Learn how to use aquaponics in plant systems. Keep chickens. Fix your own lawnmower. Fix your own toilet.

Work with your hands. Learn how to make something last without throwing it out and immediately buying a new one. Fix a chair. Learn to sew. Spend an afternoon making preserves and putting them in mason jars with your kids. Bake bread. As a nation, and as members of developed nations, we have lost our roots and the cost is going to be very high.

Additionally, people with medical knowledge are going to be in very high demand. Learn CPR. Take basic first aid training courses. If you have any calling for the medical field at all, learn. Take nursing courses. Take dental courses. Become a doula.

As the food and water resources begin to fade, people with working skill sets will survive. There will be a well paid minority of people in security or technology that move up the social ladder. Most of our population that survives, however, will do so by being prepared, and by utilizing their skills and resources properly. People that are ready and able to work hard, learn skills that have been forgotten, and bruise a few knuckles are the ones that the inherit the shining future we'll discuss in the next book.

Oh, and by the way, you might want to learn how to run a backhoe. We'll need plenty of them before it's all over.

Food – Preparing Yourself and Your Children for the Die Off

Let's take a minute and talk about food growing and food preparation. We've already talked about the scarcity of food during the die off. Even with unlimited storage space, it would be difficult to plan what your food needs are going to be, as we don't really know the length of time the crisis will last. Stored food will play a large role, no question, but you're going to need a source for fresh food as well.

Of course, during the die off, it will be difficult to have open fields filled with crops. They will be the first and most obvious target for people that have not prepared for the die off properly. As we've mentioned, there will be traditional farms as well as cattle and poultry farms - but they will be fortified.

The changing climate is also going to be a factor for outside farming, especially in the south. As temperatures go higher, less water will be available at the crucial point where more water is going to be needed for crops. You'll also see a longer season of insects and other pests.

So how do you protect your plants and animals? A greenhouse would work, but it's going to be as much as a target as an open field. What you need is a normal looking building, container, or warehouse that allows you to grow food in far less space than a traditional plot of land. Take a few moments and learn these three terms: vertical farming, hydroponic farming, and aquaponic farming.

Vertical farming is the practice of producing food in vertically stacked layers, rather than growing in farmland. Vertical farms are indoors, typically in sealed environments, so food pests aren't a problem. Right now, this makes it much easier to produce organic foods. In the future, the main advantage to indoor vertical farming is that it will allow a group of people to grow food in a building that doesn't look like an indoor garden.

Vertical farms can be started anywhere: in shipping containers, warehouses, or skyscrapers. More advanced farms use computer software with external sensors to monitor moisture levels, artificial daylight and night times, as well as growth and yield. All vertical farms produce a much larger amount of food per square foot of growing space than traditional farming.

Hydroponic farming is vertical farming, but without soil. A hydroponic farm uses nutrient-rich water to grow plants and vegetables in a substrate. These systems rely on water pumps to make sure nutrients are getting to the plant's roots, so they are susceptible to power outages. It's obviously not as cheap as planting seeds in the ground, and hydroponics requires pretty regular monitoring – but once the system is setup, you can grow large amounts of food very quickly.

Aquaponic farming is a combination of the two above types of farming, with one important addition – marine animals. With aquaponics, you don't need to add nutrients to the water. You have a tank connected to the water source with fish living in it. Very simply – the plant roots feed the fish, and the fish waste feeds the plants. Aquaponics systems can be setup with edible fish like catfish, tilapia, or bluegill. They can also be setup with koi or goldfish – but I would not recommend eating them.

Aquaponic farming uses less water than hydroponic farming and has the added bonus of providing protein as well as vegetables. Aquaponics system also requires a fair amount of maintenance and monitoring of the water and fish. However, once set up, it can feed any number of people simply by scaling up.

Any of the above three farming types can be set up in your garage, basement, or any other spare room. They are relatively easy to build and, while they do take some time, can be a hobby that may just pay off. Our changing climate and how it affects temperature and soil moisture in outdoor crops is a major factor in the rise of vertical farms. Industrial growers are aware of this and are moving forward with new designs to vertical farms, aquaponics, and hydroponics. Vertical farming produced $3 billion dollars' worth of produce in 2018 alone, and that number is expected to rise dramatically over the next few years. This means that the software, sensors, pumps, and other materials needed to invest will become better and cheaper in the coming years (and most of the software is free now).

The second factor that makes vertical farming so important is that it has to be grown in buildings, grown underground, or in mobile containers. Remember those bunkers that the military and ultra-wealthy are paying for? They'll need people to design and run their vertical farms. College campuses, fortified buildings, gated communities – when the die off comes, these people are going to need fresh produce. Having a background in vertical farming provides your children with a very real life-saving skill set when the die off comes. In addition, aquaponics has the added value of marine farming. As climate change affects our lakes and rivers, these species of fish will be harder to find in the wild. During the die off itself, land locked areas will have almost no access to fish as a protein source.

Steps You Can Take Now

If we have, at the outside, 30 years to prepare, then the average person can do a pretty good job of getting ready for the inevitable. If you are looking for a list of weapons to buy, or a checklist for digging a shelter in your backyard, you are going to be disappointed. After all, I told you in the beginning that this isn't a prepper book. However, I am going to provide you with some choices you can make that will give you a significant advantage over most of the rest of the world.

There will be people who reach this point in the book, and be angry with me about the list of "steps you can take now" that follows. After the horrors we've talked about, how can the small changes I recommend in the next few pages make any difference? My answer to this question is simple. Survivalism is a mindset. Conservation is a mindset. People cannot change from conspicuous consumerism, and become a conservationist in a day, or a week, or a year.

Consider that, as late as 100 years ago, over half of Americans lived in rural areas, which would mean that most of them WERE conservationists. Not too long before that, our country was built on several generations of survivalists. As a society, we've lost the mindset of being survivalists over the last 150 years or so. Why would we expect to be able to revert back to that mindset after reading a single book, even one with the conclusions that I've drawn here?

My advice is this – start small. Begin by making changes in your life that you feel good about. Learn some of the skills we've talked about. By just having the knowledge of the history we've talked about here, as well as the knowledge of the next die off we are facing, you already have a head start on the rest of the population.

Be prepared. Keep an open mind, and be ready to learn new things and make small changes in your life that will help you be one of the survivors. Remember the zombie apocalypse we talked about earlier? You don't have to be the fastest runner; you just don't want to be the slowest.

Above all, consider the changes you make to be positive. This is a long term vision and you are building long term goals. These changes will affect your life for the better, and will affect the lives of your children for the better. With that held firmly in mind, here is a list of steps you can take now to make sure you are one of the survivors.

1) Move out of the City. As the human die off accelerates, cities are going to become urban deserts. Any breakdown in services is going to cascade, and it will do so quickly. Electricity and water will be the first to shut down. The lack of garbage pickup will cause an explosion of rats. Without power, you can't pump gas, which means cars will be left in the streets and highways. The blockages will cause deliveries of food to come to a slowdown. What little does get through will be targeted by armed gangs.

   I'm not suggesting you buy 150 acres in the wilds of Montana, or deep in the mountains of Colorado. I'm suggesting a 5 acre homestead, with reasonable access to a small city, and within a few hours of a large one. Buy enough land to have a nice, long driveway to your home, and then, make sure and fence and gate it. It's important to have a good relationship with the neighbors (more on this in a moment), but on the other hand, maintain a level of privacy as well.

2) Outfit your home with as much "off the grid" services as possible. Again, you don't have to go crazy here. No one is suggesting a machine gun turret built in to your roof. However, understand that electricity and water service are going to be spotty at best, especially in areas not serviced by hydroelectric plants. Seriously consider drilling for water if you live in an area where oil companies aren't fracking – it doesn't cost as much as you might think. Put the pump on a battery backup system that is driven by an alternative power source. Have an extra pump and know how to maintain it yourself.

   There are three main sources of alternative energy – that is, electricity that doesn't come down the pole outside your house. The first is solar, which is expensive to get into at first, but definitely makes your money back in the long run. Wind power is also a choice, but like solar, the initial investment can scare people off. I suggest having a liquid propane or natural gas tank installed – 500 pounds would do for a short term solution – and have it attached to a small generator. Be very conservative here – you don't need to power the entire house during the crisis. Make sure to keep the fresh water coming, and any food storage needs handled. These should be your primary functions of electrical power.

Secondary, but just as important, is a method of communications – even if it's just a dedicated circuit to keep a cell phone charged, or a radio plugged in. If you can maintain a laptop or computer with an internet connection, that's ideal. Have multiple options for internet communication. If you are in a rural area, DSL or cable modems may be affected. Consider having point-to-point connection installed, getting a cellular hotspot, or installing satellite connectivity.

3) Arm yourself. If you live in a state that offers a concealed carry license, get it. Learn how to handle and fire a handgun, a shotgun, and a rifle. Teach your children to handle weapons safely. Again, no one is suggesting you go out and buy flamethrowers or a minigun. However, having a little bit of knowledge here is going to be crucial in the future.

Practice firing your weapons. Find a local gun range – I'd bet there are several indoor ranges, outdoor clay ranges, and gun clubs in your area to choose from. Maintain a store of ammunition for your weapons. Do you have to fill a room with rounds? Of course not. Most preppers and weapons experts suggest a minimum of 1000 rounds for handgun, 1000 rounds for shotgun, and 2000 rounds for rifle. Rotate your inventory as much as possible. Ammunition does have a shelf life – it's usually rated at 10 years, but if stored properly, will last much longer than that.

4) Take self-defense classes. They are fun, they are a great workout, and they are exceptionally empowering, especially for women and children. No one is suggesting you become the next Bruce Lee, but a working knowledge of defense movement will be an incredible advantage over most of the population. I'm not saying to go out and get a black belt in karate. I'm saying that your local community center, the YMCA, or your local fitness center will all have basic classes available that teach defensive movement.

   Become the example to your children or the other significant people in your life. Rotate around – there's no need to follow one discipline. Krav Maga, Jiu jitsu, and kickboxing are all great disciplines and, if taught well, brings a lifetime of fitness and self-confidence to the people that practice them.

5) Train yourself to keep rotating stock of medical supplies and other important consumables. How many people can say they have a well-stocked medicine cabinet? I'm not talking about band-aids or sunburn cream, I'm talking about a small trauma kit with alcohol, gauze, and medical tape. You can buy a great trauma kit for less than $100 for the house, and smaller ones for the car. Make sure the kit has trauma shears, a CPR mask, and clotting sponges in addition to gauze and medical wrap. Don't know how to use them? Take a weekend first aid class. Your instructor, if any good at all, will be thrilled to show you how to use these medical supplies.

Make it a habit to get extras when you go to the store. Need to buy a bottle of isopropyl alcohol? Get an extra. Keep an extra toothbrush. Buy an extra bag of charcoal or bottle of propane for the grill. Make sure and keep candles with a few lighters. Keep a flashlight handy and in good working order, with extra batteries in a place you can find them.

Buy and store an emergency food supply bucket, or a supply of MREs. Food supply buckets have multiple freeze-dried meals in them, with a 25 year shelf life. A decent sized, sealed bucket costs less than a dinner out. MRE's have a shorter shelf life, but tend to be a little tastier. Keep either in the bottom of your pantry, your closet, or with camping supplies. Keep a rotating stock of extra canned goods. Keep a small store of cooking fuel.

6) If you have enough land, a fenced yard, and a little extra time, get some animals. You don't need to put a commercial chicken processing plant on your land, but a few chickens or geese are a lot of fun, especially if you have kids, and they are exceptionally easy to maintain. There are other advantages – chickens keep your yard and gardens pest free without using poison, and they produce eggs. Goats are a huge amount of fun and keep your yard trimmed and fertilized all at the same time.

7) Know your neighbors. This can be as easy as having an annual Christmas party, or spring BBQ. You don't have to talk about the end of civilization as we know it – trust me, it makes conversation awkward. A working knowledge of the people that live close to you is a good thing. It invokes a sense of community, allows people to bond over common beliefs, and takes communication out of cyberspace and into the real

world. From a practical standpoint, identifying community leaders, knowing names and kids names, hobbies, and what your neighbors do for a living will be very helpful when the die off comes.

8) Learn how to grow, cook, and store food. As a parent myself, it's unbelievable to me how many high school students, college students, and young adults don't have the first clue about cooking, storing food, or even the basics of food budgeting. A recent case that made worldwide headlines was of a 20 year old college student who ate 5 day old pasta and died after his liver shut down. Reported in Belgium, the student had contracted food poisoning from a relatively common bacterium associated with leaving fried rice or pasta at room temperature for days and then consuming it.

Here's a few things that you can do now with your kids. Note that these are not long term plans that will make your child the next leader of the Resistance. However, these projects are a way for children to get into the mindset of sourcing healthy food, in any circumstance. The idea here is to give your children a basis to start from. These lessons and suggestions, while seemingly innocuous, will give them a huge advantage over their peers.

- Take your children with you to the grocery store. Encourage them to take their own cart and get their own items that they want for meal sides and snacks. At first, you'll get cookies and chips, but you'll be surprised at how quickly those items lose their draw when they are readily available. Focus on fresh fruits, nuts, and vegetables. If your kid loves hot pockets, then that's fine. Can a child get the ingredients and recipes online to make their own hot pockets? Absolutely. And they taste better, too.

- Cook with your children. Work with them on recipe choices, meal preparation, and how to read cookbooks and interpret measurements. Work with your kids on oven and stovetop safety. Try recipes with locally grown produce. Talk about basic nutrition with your kids. Remember that bananas, mangos, and other tropical fruit won't be available during the die off. What local vegetables are high in Vitamin C? What other types of protein are there?

- Plant a garden with your kids. Do you have to plant an acre of garden vegetables? Of course not. However, having a basic knowledge of plants and soil will be a major advantage in the future. Learn about nitrogen-based plant foods and fertilizers and how they help the soil. Get your hands dirty in containers. Learn alternative methods of planting, such as upside-down tomato planters and raised beds. Tomatoes, herbs, and peppers are all easy to grow and can be used for daily meals. Growing plants and vegetables is fun, and it's a great way to bond with your kids as well.

- Food storage. Spend an afternoon canning food with your kids. Learn how to make pickles, preserves, jams, and jelly. These are fun, easy projects that don't cost a lot of money and have a lifetime of rewards in knowledge. In addition, you control what goes into the food your kids are eating. The added bonus is your child gets to pull out what was made 3 months later, or 6 months later, or a year later, and have a positive memory of the time you spent with them.

If there is one remaining step you can take now, it's to educate yourself. I implore you to take a few minutes each day, or an afternoon on the weekends and learn a new skill. This doesn't involve anything superhuman. It's not a major lifestyle change, and you don't have to walk around with a sign shouting "The End is Near!". Take a CPR class. Go camping. Join a shooting club.

Most of the ways you can begin to prepare yourself and your children for the coming die off are basic changes that can enhance your skillset. Learn how to fix things instead of just throwing them away and buying a new one. Disconnect from your phone and your other devices and read a book. Watch the occasional documentary. Learn how to do things with your hands. Learn a skill. Play an instrument. Write a book. Build a birdhouse. Plant a garden. After all, without these things, what are we living for, anyway?

Keeping Your Loved Ones Safe During the First Year

In most of human history and pre-history, humans gathered together to fight a common enemy. They built walls to repel the enemy. They built homes to protect themselves and their food stores from animals. They forged alliances with their neighbors to increase their defenses in case of attack by an aggressor. Alliances also allowed for a larger and more diverse labor pool. Most importantly, alliances allowed a group of like-minded people to procreate and expand their territory.

I think the next human die off will have a relatively short duration – no more than five years, and probably less than two years from the beginning of the crisis to the point that we are able to bury the bodies and begin restructuring. It's difficult to discuss long term practices for safety because the variables that cause the tipping point are still unknown. We've discussed possible scenarios, but the truth is that there are a variety of factors, from climate change, to geopolitical tensions, to water and food availability that will affect the length and severity of the tipping point and the die off itself. It's these variables that will determine the length of the crisis.

If the crisis lasts longer than a few years, alliances with your neighbors and other survivors will become inevitable. No matter how well you plan, individual resources will eventually run out. Human beings are built to depend on one another – we require on a very primal level the acceptance of other like-minded people. It's part of the reason we are the dominant life form on the planet today.

Of course, by year two after the tipping point, most of the people that haven't prepared will be dead anyway. The first year will be the most tumultuous, and the most dangerous as unprepared people become desperate. In the first year, once the tipping point is reached and the die off begins, how can you keep your family safe?

There will be fortified campuses for wealthy and well-trained people. There will still be existing, but heavily guarded, cattle and produce farms. There will be distribution warehouses that are heavily guarded, and filled with goods that are delivered to these campuses by drone and resupplied by armed convoys. Underground bunkers will be sealed, and the ruling class will wait for the time when they can re-open the blast doors and attempt to retake control of the country. If you are one of these minority, I salute you.

If, however, you are one of the other 99% of the population living in a developed country today, you have to assume that there is no one else coming to your rescue, at least for the first year. Law enforcement is going to be in complete shambles in urban centers, and completely non-existent in rural areas. In the best-case scenario, the armed forces will be tasked with defending our country, at worst case, defending the bunkers that the ruling class will be hiding in. At no time will the military be used to protect individuals. Make no mistake – you will be on your own.

The first year of the crisis will be the most dangerous. Breakdowns in authority and law enforcement will cause riots, hoarding, price gouging, and revolts. You must be prepared to take the lessons learned and the resources that you have saved and put them to good use, and quickly. Once the tipping point is reached and the die off begins, you will be forced to deal with circumstances that will change rapidly. You must be prepared, at this point, to accept that the resources you have carefully put aside, and the skills you have learned, are now your lifeline and the sole means of keeping you and your loved ones alive.

The first rule during the crisis will be the hardest to follow. You cannot help anyone that hasn't prepared for the die off. Doing so puts yourself and your family at risk. There will be people dying all around you of starvation, people who will beg you for the sake of their own children to help them with food, water, or fuel. Any assistance you give them is a drain on your resources, and only prolongs their agony. YOU CANNOT HELP THEM. Doing so only increases your own risk.

If at all possible, stay home. If you are in a city that has been the victim of an attack, or is expecting an impending natural disaster, then obviously, that's a cause to head out. But DO NOT throw a hastily packed two-day suitcase in your car and join a huge crowd of evacuees. Nothing will get you and your family killed faster than joining the herd mentality. Stay at home and collect the available resources you have there. Even if you haven't planned for the coming die off, you will be in much better shape at home than on the road.

Keep a low profile. There's no need to be flashy with weapons or food. Parading your sustainability will only make you a target for people that have not prepared properly. Resist the urge to say "I told you so" to people that come to your door looking for food or other supplies. Keep your inventory of supplies to yourself.

Be prepared to defend yourself, your family, and your property. Be prepared to kill if necessary to protect your resources for the sake of your survival, and the survival of the people that depend on you. Be ready and mentally prepared to shoot first and without hesitation or remorse. Do NOT shoot to wound or to warn. Shoot to kill. It will be absolutely necessary for you to have the ability and the mental strength to shoot first, to aim accurately, to drop the person who is attempting to harm you or your family, and to do all three of these without a second thought.

Maintain your connection to the rest of the world. If you've prepared properly, you'll have internet connection or a cellular connection to communicate and receive information about the crisis. Don't listen to opinions or third-party retelling. Get the facts yourself.

Put your knowledge of your neighbors to good use. By now, you should know who has resources or skills that can be traded for. Be aware that depending on your location or circumstances, you may need to set guards on areas of ingress and egress to your neighborhood, especially if you live directly off a major freeway. You may have to dig trenches in roads, or isolate areas using cars to block roadways. Know who you can trust among your neighbors to be a constructive asset during this first year. Know who among your neighbors is the leader. If you look around and no one else is the leader, then it's probably you.

Do not let your children out of your sight during the crisis, not even to people you trust. By the nature of our humanity, our children are our weak point. They are the "Achilles heel" in your defenses. Do not weaken your position by allowing them to go somewhere without you, at least not for the first year.

Do not accept cash for anything. Banks will survive the die off, but not the conventional brick-and-mortar bank we recognize now. All transactions will happen online. Drone deliveries will be paid for online. Wealth will not be measured in paper money during or after the die off, indeed, it barely is now. Do NOT make the mistake of selling your supplies for paper money, no matter what the dollar amount offered is.

Be flexible and keep your mind open to alternative solutions. You simply can't prepare for every eventuality – it's not possible, and you'll go crazy trying. However, if you've prepared yourself properly, you can weather any storm. Stay positive. Cry by yourself if you have to, but be strong and stay upbeat for the people that are relying on you. Every problem has a solution. Be willing to accept that fact and come up with creative solutions for the problems that you will surely face.

Above all, take care of your children and your loved ones. They are your immortality. Teach them and guide them through the crisis. There is no doubt that the coming die off will easily be the darkest time in our recorded history. Hopefully, if you've read this book to this final page, you'll know that there is hope for humanity after the die off. The coming human die off isn't the end. It isn't even the beginning of the end. It is the end of the beginning of the next stage of our evolution. Be ready. Prepare. Live to see the future. It will be worth living for.

Coming soon – Book 2 – The Shining Future.

Afterword

One of my favorite songs by The Who has a line in it that says: "Meet the new boss, same as the old boss". The song title is "Won't Get Fooled Again", and most people born after 1971 will recognize it as the theme song to a popular television crime drama. Anyone who actually lived through the 60's will recognize the song for what it really was: a call to action for violent overthrow of the government.

The line (one of the most famous of all The Who lyrics) means that nothing ever changes. However, I don't believe that will be the case after the die off. I don't believe that things will return to "business as usual". I think that our mindset as a collective race — a race of humans — will change so dramatically as we emerge from the crisis that I have described in these pages as to be unrecognizable to who and what we are now. I believe this change will be as drastic, as violent, and ultimately, as beautiful as a caterpillar changing into a butterfly.

The road to this next stage in our development as a species will be difficult and dark. As an American, I fear for my homeland the most, as I think anyone from any country would do. America and its residents have labored under a suspension of belief for so many years that I don't think anything less than a catastrophic event can change the mindset of Americans and our elected leaders. There are so many issues facing us that it's hard to differentiate and prioritize which ones are most important and which ones we can let go of. This plethora of problems — this overage of issues — has deadlocked the American public into believing that change on any major issue is impossible. There are notable exceptions, sure: the legal acceptance of gay marriage comes to mind. However, I think that the majority of Americans are living with little or no hope in any measurable change.

The simple fact of the matter is that change is inevitable. The cyclic nature of world economies, our planet, and our race as a whole has ridden the waves of change for thousands of years. Nations rise and fall. Populations grow and shrink. Some of these changes are tidal, in that they can be measured according to a predetermined cycle. Some changes are not on a measurable timeline. But make no mistake – "the only true constant in life is change".

One of the most difficult things about writing a book on current events that at some point, you have to stop writing and publish. The biggest news from 2020 has been, of course, the coronavirus and the disease it causes in humans, called COVID-19. This pandemic is still developing as I write this, and the global effects of this modern disease are going to be felt for years. However, what's interesting about the coronavirus is the mortality rate. Because of the development and use of anti-viral drugs, the mortality rate of COVID-19 is between 2% and 3%. This means, that for the first time in history, we as a species have been able to significantly lessen the mortality rate of a pandemic virus.

There are several things I still want to tell you about, and if you like my unique viewpoints on the state of the world and climate change, then you'll have to subscribe to my online newsletter, read my blog posts, or hire me to speak. Doing any of these will also be a great way to find out the release date of my next book.

The good news about the topic of climate change (at least from the standpoint of this book) is that we are racing toward a climate disaster, and no one seems to be paying any attention. Our attitudes aren't changing, which means something else must. The inevitable conclusion to the misuse of our planet's resources is a human die off. I suppose that in this way, my book becomes more relevant with each passing day. Every day that we do not take action is another step towards the mass grave I've predicted in this book. It IS coming and only the people who have prepared for it will survive.

Appendix A – The Green New Deal

H. RES. 109

IN THE HOUSE OF REPRESENTATIVES
February 7, 2019

RESOLUTION

Recognizing the duty of the Federal Government to create a Green New Deal.

Whereas the October 2018 report entitled "Special Report on Global Warming of 1.5 ºC" by the Intergovernmental Panel on Climate Change and the November 2018 Fourth National Climate Assessment report found that—

(1) human activity is the dominant cause of observed climate change over the past century;

(2) a changing climate is causing sea levels to rise and an increase in wildfires, severe storms, droughts, and other extreme weather events that threaten human life, healthy communities, and critical infrastructure;

(3) global warming at or above 2 degrees Celsius beyond preindustrialized levels will cause—

(A) mass migration from the regions most affected by climate change;

(B) more than $500,000,000,000 in lost annual economic output in the United States by the year 2100;

(C) wildfires that, by 2050, will annually burn at least twice as much forest area in the western United States than was typically burned by wildfires in the years preceding 2019;

(D) a loss of more than 99 percent of all coral reefs on Earth;

(E) more than 350,000,000 more people to be exposed globally to deadly heat stress by 2050; and

(F) a risk of damage to $1,000,000,000,000 of public infrastructure and coastal real estate in the United States; and

(4) global temperatures must be kept below 1.5 degrees Celsius above preindustrialized levels to avoid the most severe impacts of a changing climate, which will require—

(A) global reductions in greenhouse gas emissions from human sources of 40 to 60 percent from 2010 levels by 2030; and

(B) net-zero global emissions by 2050;

Whereas, because the United States has historically been responsible for a disproportionate amount of greenhouse gas emissions, having emitted 20 percent of global greenhouse gas emissions through 2014, and has a high technological capacity, the United States must take a leading role in reducing emissions through economic transformation;

Whereas the United States is currently experiencing several related crises, with—

(1) life expectancy declining while basic needs, such as clean air, clean water, healthy food, and adequate health care, housing, transportation, and education, are inaccessible to a significant portion of the United States population;

(2) a 4-decade trend of wage stagnation, deindustrialization, and antilabor policies that has led to—

(A) hourly wages overall stagnating since the 1970s despite increased worker productivity;

(B) the third-worst level of socioeconomic mobility in the developed world before the Great Recession;

(C) the erosion of the earning and bargaining power of workers in the United States; and

(D) inadequate resources for public sector workers to confront the challenges of climate change at local, State, and Federal levels; and

(3) the greatest income inequality since the 1920s, with—

(A) the top 1 percent of earners accruing 91 percent of gains in the first few years of economic recovery after the Great Recession;

(B) a large racial wealth divide amounting to a difference of 20 times more wealth between the average white family and the average black family; and

(C) a gender earnings gap that results in women earning approximately 80 percent as much as men, at the median;

Whereas climate change, pollution, and environmental destruction have exacerbated systemic racial, regional, social, environmental, and economic injustices (referred to in this preamble as "systemic injustices") by disproportionately affecting indigenous peoples, communities of color, migrant communities, deindustrialized communities, depopulated rural communities, the poor, low-income workers, women, the elderly, the unhoused, people with disabilities, and youth (referred to in this preamble as "frontline and vulnerable communities");

Whereas, climate change constitutes a direct threat to the national security of the United States—

(1) by impacting the economic, environmental, and social stability of countries and communities around the world; and

(2) by acting as a threat multiplier;

Whereas the Federal Government-led mobilizations during World War II and the New Deal created the greatest middle class that the United States has ever seen, but many members of frontline and vulnerable communities were excluded from many of the economic and societal benefits of those mobilizations; and

Whereas the House of Representatives recognizes that a new national, social, industrial, and economic mobilization on a scale not seen since World War II and the New Deal era is a historic opportunity—

(1) to create millions of good, high-wage jobs in the United States;

(2) to provide unprecedented levels of prosperity and economic security for all people of the United States; and

(3) to counteract systemic injustices: Now, therefore, be it

Resolved, That it is the sense of the House of Representatives that—

(1) it is the duty of the Federal Government to create a Green New Deal—

(A) to achieve net-zero greenhouse gas emissions through a fair and just transition for all communities and workers;

(B) to create millions of good, high-wage jobs and ensure prosperity and economic security for all people of the United States;

(C) to invest in the infrastructure and industry of the United States to sustainably meet the challenges of the 21st century;

(D) to secure for all people of the United States for generations to come—

(i) clean air and water;

(ii) climate and community resiliency;

(iii) healthy food;

(iv) access to nature; and

(v) a sustainable environment; and

(E) to promote justice and equity by stopping current, preventing future, and repairing historic oppression of indigenous peoples, communities of color, migrant communities, deindustrialized communities, depopulated rural communities, the poor, low-income workers, women, the elderly, the unhoused, people with disabilities, and youth (referred to in this resolution as "frontline and vulnerable communities");

(2) the goals described in subparagraphs (A) through (E) of paragraph (1) (referred to in this resolution as the "Green New Deal goals") should be accomplished through a 10-year national mobilization (referred to in this resolution as the "Green New

Deal mobilization") that will require the following goals and projects—

(A) building resiliency against climate change-related disasters, such as extreme weather, including by leveraging funding and providing investments for community-defined projects and strategies;

(B) repairing and upgrading the infrastructure in the United States, including—

(i) by eliminating pollution and greenhouse gas emissions as much as technologically feasible;

(ii) by guaranteeing universal access to clean water;

(iii) by reducing the risks posed by climate impacts; and

(iv) by ensuring that any infrastructure bill considered by Congress addresses climate change;

(C) meeting 100 percent of the power demand in the United States through clean, renewable, and zero-emission energy sources, including—

(i) by dramatically expanding and upgrading renewable power sources; and

(ii) by deploying new capacity;

(D) building or upgrading to energy-efficient, distributed, and "smart" power grids, and ensuring affordable access to electricity;

(E) upgrading all existing buildings in the United States and building new buildings to achieve maximum energy efficiency, water efficiency, safety, affordability, comfort, and durability, including through electrification;

(F) spurring massive growth in clean manufacturing in the United States and removing pollution and greenhouse gas emissions from manufacturing and industry as much as is technologically feasible, including by expanding renewable energy manufacturing and investing in existing manufacturing and industry;

(G) working collaboratively with farmers and ranchers in the United States to remove pollution and greenhouse gas emissions from the agricultural sector as much as is technologically feasible, including—

(i) by supporting family farming;

(ii) by investing in sustainable farming and land use practices that increase soil health; and

(iii) by building a more sustainable food system that ensures universal access to healthy food;

(H) overhauling transportation systems in the United States to remove pollution and greenhouse gas emissions from the transportation sector as much as is technologically feasible, including through investment in—

(i) zero-emission vehicle infrastructure and manufacturing;

(ii) clean, affordable, and accessible public transit; and

(iii) high-speed rail;

(I) mitigating and managing the long-term adverse health, economic, and other effects of pollution and climate change, including by providing funding for community-defined projects and strategies;

(J) removing greenhouse gases from the atmosphere and reducing pollution by restoring natural ecosystems through proven low-tech solutions that increase soil carbon storage, such as land preservation and afforestation;

(K) restoring and protecting threatened, endangered, and fragile ecosystems through locally appropriate and science-based projects that enhance biodiversity and support climate resiliency;

(L) cleaning up existing hazardous waste and abandoned sites, ensuring economic development and sustainability on those sites;

(M) identifying other emission and pollution sources and creating solutions to remove them; and

(N) promoting the international exchange of technology, expertise, products, funding, and services, with the aim of making the United States the international leader on climate action, and to help other countries achieve a Green New Deal;

(3) a Green New Deal must be developed through transparent and inclusive consultation, collaboration, and partnership with frontline and vulnerable communities, labor unions, worker cooperatives, civil society groups, academia, and businesses; and

(4) to achieve the Green New Deal goals and mobilization, a Green New Deal will require the following goals and projects—

(A) providing and leveraging, in a way that ensures that the public receives appropriate ownership stakes and returns on investment, adequate capital (including through community grants, public banks, and other public financing), technical expertise, supporting policies, and other forms of assistance to communities, organizations, Federal, State, and local government agencies, and businesses working on the Green New Deal mobilization;

(B) ensuring that the Federal Government takes into account the complete environmental and social costs and impacts of emissions through—

(i) existing laws;

(ii) new policies and programs; and

(iii) ensuring that frontline and vulnerable communities shall not be adversely affected;

(C) providing resources, training, and high-quality education, including higher education, to all people of the United States, with a focus on frontline and vulnerable communities, so that all people of the United States may be full and equal participants in the Green New Deal mobilization;

(D) making public investments in the research and development of new clean and renewable energy technologies and industries;

(E) directing investments to spur economic development, deepen and diversify industry and business in local and regional economies, and build wealth and community ownership, while prioritizing high-quality job creation and economic, social, and environmental benefits in frontline and vulnerable communities, and deindustrialized communities, that may otherwise struggle with the transition away from greenhouse gas intensive industries;

(F) ensuring the use of democratic and participatory processes that are inclusive of and led by frontline and vulnerable communities and workers to plan, implement, and administer the Green New Deal mobilization at the local level;

(G) ensuring that the Green New Deal mobilization creates high-quality union jobs that pay prevailing wages, hires local workers, offers training and advancement opportunities, and guarantees wage and benefit parity for workers affected by the transition;

(H) guaranteeing a job with a family-sustaining wage, adequate family and medical leave, paid vacations, and retirement security to all people of the United States;

(I) strengthening and protecting the right of all workers to organize, unionize, and collectively bargain free of coercion, intimidation, and harassment;

(J) strengthening and enforcing labor, workplace health and safety, antidiscrimination, and wage and hour standards across all employers, industries, and sectors;

(K) enacting and enforcing trade rules, procurement standards, and border adjustments with strong labor and environmental protections—

(i) to stop the transfer of jobs and pollution overseas; and

(ii) to grow domestic manufacturing in the United States;

(L) ensuring that public lands, waters, and oceans are protected and that eminent domain is not abused;

(M) obtaining the free, prior, and informed consent of indigenous peoples for all decisions that affect indigenous peoples and their traditional territories, honoring all treaties and agreements with indigenous peoples, and protecting and enforcing the sovereignty and land rights of indigenous peoples;

(N) ensuring a commercial environment where every businessperson is free from unfair competition and domination by domestic or international monopolies; and

(O) providing all people of the United States with—

(i) high-quality health care;

(ii) affordable, safe, and adequate housing;

(iii) economic security; and

(iv) clean water, clean air, healthy and affordable food, and access to nature.

Appendix B – Books I've Read That Have Helped Me Write This Book

The 6th Extinction: An Unnatural History
By Elizabeth Kolbert

Singularity Rising
By James D. Miller

Brave New World
By Aldous Huxley

Survive Anything
By Beau Griffin

Profit Over People
By Noam Chomsky

Pandemic 1918
By Catharine Arnold

40 Projects for Building Your Backyard Homestead
By David Toht

The Time Machine
By H. G. Wells

Why Nations Fail
By Daron Acemoglu and James A. Robinson

The Singularity is Near
By Ray Kurweil

Losing Earth – A Recent History
By Nathaniel Rich

Homo Deus – A Brief History of Tomorrow
By Yuval Noah Harari

An Inconvenient Truth
By Al Gore

The Uninhabitable Earth
By David Wallace-Wells

1776
By David McCullough

The Stand
By Stephen King

www.ingramcontent.com/pod-product-compliance
Lightning Source LLC
Chambersburg PA
CBHW030616220526
45463CB00004B/1304